Relaxation in Physical and Mechanical Behavior of Polymers

Relaxation in Physical and Mechanical Behavior of Polymers

Maxim Arzhakov

CISP

CRC Press
Taylor & Francis Group
Boca Raton London New York

CRC Press is an imprint of the
Taylor & Francis Group, an **informa** business

Translated from Russian by V.E. Riecansky

CRC Press
Taylor & Francis Group
6000 Broken Sound Parkway NW, Suite 300
Boca Raton, FL 33487-2742

First issued in paperback 2021

© 2019 by CISP
CRC Press is an imprint of Taylor & Francis Group, an Informa business

No claim to original U.S. Government works

ISBN 13: 978-1-03-223737-4 (pbk)
ISBN 13: 978-0-367-19982-1 (hbk)

Contents

Introduction

In everyday practice, the physical bodies surrounding us exist in three aggregate states, namely, solid, liquid and gaseous[1], realized with the appropriate combination of temperature and pressure. At atmospheric pressure, the only factor determining the existence of a given physical body in one or another aggregate state is temperature. The solid is converted to liquid at the melting point for crystalline substances or at a flow temperature – for amorphous. At the boiling point, the liquid passes into a gaseous state.

In the conditions of gravity, i.e. under the influence of its own weight, the solid body retains both volume and shape, the liquid retains the volume but not the shape, and the gas does not retain either one or the other. From molecular–kinetic positions, these effects are dictated by the energy ratio of intermolecular or interatomic interactions U and thermal energy kT.

For solid bodies, $U \gg kT$, which causes the absence of translational translations or mutual displacements of particles relative to each other. It is this factor that explains the constancy of the volume and shape characteristic of a given aggregate state. In liquids, $U \sim kT$. At a given ratio of the energy of intermolecular interactions, it is still sufficient to keep the liquid in a condensed state. However, in this case, the progressive degrees of freedom are already 'unfrozen' for the particles, as a result of which the fluid flows under the influence of its own weight and assumes the shape of a vessel, preserving the volume. For gases $U \ll kT$, the energy of intermolecular interactions is insufficient to ensure the existence of a stable condensed state, and as a result, the gas fills all available space, i.e. does not save either volume or shape.

Thus, we can conclude that this substance in a given aggregate state is characterized by a certain behaviour, which causes the

---

[1]At present, the fourth aggregate state includes plasma.

formation of a certain set of properties. The situation, however, is significantly complicated when taking into account the time factor.

Thus, for example, natural resins retain a volume and shape at small observation times (minutes, hours), i.e. formally refer to solid objects. When the observation time increases to weeks and months, liquid-like fluidity, accompanied by loss of form, appears in them. Hurricane wind breaks trees and destroys buildings because at a speed of more than 100 km/h, the air becomes stiffer than typical hard materials such as wood and stone. In shock, almost instantaneous influences, the gas behaves as a liquid, and the liquid fractures by brittle fracture like a solid body. In other words, a physical body in a certain aggregate state is able to demonstrate a complex of properties that is characteristic of another aggregate state, depending on the time or rate of impact. Such processes, the course and result of which depend on time, are referred to as relaxation, and the phenomenon itself is referred to as relaxation.

In physics, relaxation (from Latin *relaxatio* – weakening, reduction) is interpreted as the process of establishing thermodynamic (statistical) equilibrium in a system consisting of a large number of particles. In this interpretation, the classical relaxation theory is most fully developed for gases taking into account both the translational and internal degrees of freedom of a given class of substances.

In a more general and, perhaps, simplified version, relaxation is considered as the response of the system to an external action. Such a macroscopic reaction is due to microscopic displacements of the kinetic units composing the system[2], where the reaction time or the relaxation time of the system is determined by the τ time necessary for these microscopic rearrangements. Obviously, the relaxation time is inversely proportional to temperature – the higher the temperature, the more intense the thermal motion of the kinetic units, the higher the speed of their mutual displacement and less than τ.

The very fact of the course of the relaxation process, and, consequently, the character of the reaction of the system depends on the ratio of the relaxation time τ and the time of the external action t. The solid reaction of the system is observed when $t \ll \tau$. For such a short time, the effects of the rearrangement of the kinetic units do not have time to occur, and the object exhibits an elastic response with a tendency to brittle failure. For liquids and, the more so, gases, the ratio $t \gg \tau$ is characteristic. In these conditions, the

---

[2]Kinetic unit – a structural unit that has translational and vibrational degrees of freedom, which provides the processes of mass transfer in the system.

system has time to adequately restructure or 'relax' due to mutual displacements of kinetic units.

From these positions, the above examples become fully understandable. Under isothermal conditions for a sample of natural resin, the time necessary for the rearrangement of the kinetic units or the relaxation time is constant. If the observation time or, which is the same, the time of the action of the own weight is small ($t \ll \tau$), mutual translational motion of the kinetic units does not have time to occur, the object preserves its form, which allows us to identify it as a solid body. As the observation time increases, the relation $t \gg \tau$, which 'resolves' the translational rearrangements of the kinetic units, is realized, which determines the macroscopic liquid-like behaviour of the sample, namely, its flow without conservation of shape.

With a moderate wind speed, the molecules of the gases entering into its composition manage to 'swirl' the obstacle, without causing any damage to it. With an increase in the speed of time, such rearrangements become insufficient, and the effect of wind on any object becomes comparable in strength to the impact of an elastic solid body.

In the application to various aspects of physics and physical chemistry of condensed polymeric substances and materials, the concept of 'relaxation' has, on the one hand, a broader meaning, and on the other – acquires a number of specific features and characteristics.

In the context of the proposed monograph, relaxation refers to changes in the structural and physical state of a polymer body and its complex of properties that occur in time and are caused by the establishment in the system of statistical equilibrium disturbed by external action – temperature, mechanical, electrical, magnetic, acoustic, etc. We specify the key concepts of this definition in terms of the physical mechanics of polymers, the questions of which are given the most attention in this work.

First, the development of the physico-mechanical process involves the perturbation of the system by the action of thermal and mechanical (force) fields and the reaction to perturbation, i.e. the tendency to return to an equilibrium state or proper relaxation under given conditions. The result of the elementary act of 'disturbance – relaxation' is largely determined by the regime of temperature–force action. In modern instrumental practice, such regimes include

- static, when a constant load or deformation under isothermal conditions is applied to the sample;

- dynamic, when the temperature-force parameters of the external disturbance vary with a constant speed, according to the harmonic law, or impulsively.

Secondly, the observed relaxation phenomena are due to mutual displacements of the kinetic units with the overcoming of the potential barrier. For polymers, the set of such kinetic units is extremely wide and includes

- side groups of the macromolecule;
- length of chain, less than segment;
- segments;
- macromolecules;
- supersegmental and supramolecular structural formations.

In this case, these movements may be accompanied by conformational transitions, for example, from twisted conformation to rectified and vice versa, as well as structural rearrangements with the destruction of the initial and the formation of a new type of crystal structure, pre-crystallization, recrystallization, polymorphic transformations, secondary phase transitions, and the like.

Thirdly, the rate of development of the physico-mechanical process is determined by the time necessary for the above-mentioned rearrangements of the kinetic units, or by the relaxation time of a particular kinetic unit. A wide range of kinetic units of different sizes dictate the broad spectrum of relaxation times characteristic of polymers in the interval from picoseconds to decades.

Fourth, moving large-scale (starting from a segment) kinetic units in addition to the thermal energy necessary to overcome a potential barrier requires the presence of voids in the neighbourhood, i.e. a certain fraction of the free volume.

Fifthly, relaxation as a tendency to establish a statistical equilibrium in the polymeric body due to microscopic rearrangements involving kinetic units is macroscopically manifest

- in time dependencies of various mechanical characteristics of the material (mechanical stress, strain, modulus of elasticity, etc.) in the case of a static exposure regime;
- in the dependences of physical-mechanical parameters (modulus of elasticity, limit of stimulated elasticity, glass transition temperature and melting, etc.) on the speed or frequency of the effect in the case of dynamic regimes.

Without claiming a comprehensive coverage of the problem, the author set himself the goal of revealing the role of relaxation phenomena that determines the viscoelasticity and plasticity of polymers, the specifics of their melting and crystallization, the behaviour of deformed polymeric materials, and the optimal modes of processing thermoplastics.

The book is equipped with a number of applications that detail the methodology for studying relaxation phenomena, the features of the structural organization of amorphous and semicrystalline polymers, as well as approaches to the unified description of deformation processes that result from the relaxation nature of the physico-mechanical behaviour of materials.

Viscoelasticity and relaxation

The mechanical properties of physical bodies, including polymers, determine their response to external force.

Under the *stress* $\sigma = F/S_{c.s.}$, where F is the applied force, $S_{c.s.}$ is the cross-sectional area of the sample, the physical body is deformed, changing the shape and dimensions.

The amount of *strain* is estimated as the relative change in the body size. For example, in the case of uniaxial tension, deformation is defined as $\varepsilon = \dfrac{l - l_0}{l_0}$, where l is the linear dimension of the deformed body; l_0 is the original linear dimension.

The relationship between stress and strain is described by two basic laws:

for ideal elastic bodies – *Hooke's law*

$$\sigma = E\varepsilon, \tag{1.1}$$

where E is the elastic modulus or Young's modulus.

for ideal liquids – *Newton's law*

$$\sigma = \eta \frac{d\varepsilon}{dt}, \tag{1.2}$$

where η is the viscosity of the liquid; $d\varepsilon/dt$ is the strain rate or the velocity of the viscous flow.

In the vast majority of cases, *the behaviour of real condensed bodies, especially polymers, is a combination of elasticity and viscous flow. This behaviour is denoted as viscoelasticity, and similar bodies are called viscoelastic.*

The relaxation nature of viscoelasticity manifests itself as a dependence of the mechanical behaviour of the material on the time of observation or on the time of the force action.

For example, if you watch a piece of bitumen for an hour, then during this period of time it will completely retain its volume and shape, which is a sign of a solid. The flow of bitumen at room temperature under the influence of its own weight, i.e. liquid-like behaviour while maintaining volume, but not shape, becomes noticeable only when the observation time increases to several weeks. The same bitumen under impact loading demonstrates a response typical of an elastic solid and is described by Hooke's law. When force is applied to the sample for a long time, bitumen flows like a viscous liquid, in accordance with Newton's law.

Mechanical models have been proposed back in the XIX century to explain the time-dependent response of viscoelastic natural resins (tar, bitumen, etc.) and their behaviour is discussed below.

1.1. Mechanical models of a viscoelastic body

The basic elements of mechanical models are a spring and a piston in a viscous liquid [1–7].

The deformation of a spring with elastic modulus E obeys Hooke's law (expression (1.1)), and when unloading the spring elastically relaxes, returning to its original state, i.e. taking the original dimensions. In other words, in the 'load–unload' mode, the residual deformation of the spring ε_{res} is equal to zero ($\varepsilon_{res} = 0$).

The motion of the piston in a liquid medium with viscosity η obeys Newton's law (expression (1.2)). When unloading, the piston does not return to the initial position, and the residual deformation of the system ε_{res} is equal to the applied deformation ε ($\varepsilon_{res} = \varepsilon$).

The Maxwell mechanical model is a series-connected elastic element (spring with elastic modulus E) and a viscous element (a piston in a liquid with a viscosity of η).

In tension, the general deformation of the Maxwell model consists of the elastic strain of the spring ε_{el} and the deformation ε_{η} caused by the motion of the piston in a viscous medium:

$$\varepsilon = \varepsilon_{el} + \varepsilon_{\eta}. \tag{1.3}$$

After differentiation and the corresponding substitution, taking into account the Hooke and Newton laws, we obtain

$$\frac{d\varepsilon}{dt} = \frac{1}{E}\frac{d\sigma}{dt} + \frac{\sigma}{\eta} \tag{1.4a}$$

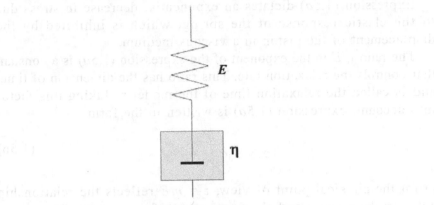

Maxwell's model.

or

$$\frac{d\sigma}{dt} = E\frac{d\varepsilon}{dt} - \frac{\sigma}{\eta / E}. \qquad (1.4b)$$

With instantaneous stretching to a given amount of strain ε, an instantaneous stress σ_0 arises in the model, caused by deformation of the elastic spring. If the given deformation is kept constant, the spring tends to resiliently return to its original state. This requires the displacement of the piston in a viscous fluid, which determines the slow nature of the relaxation of the model. This behaviour should be monitored by the time dependence $\sigma = f(t)$ (Fig. 1.1a) or, in other words, by **stress relaxation**.

Under the conditions of the stress relaxation experiment $\varepsilon = $ const and $d\varepsilon/dt = 0$. In this case, expression (1.4b) is written as

$$\frac{d\sigma}{dt} = -\frac{\sigma}{\eta / E}.$$

After integration we obtain

$$\ln\frac{\sigma_t}{\sigma_0} = -\frac{t}{\eta / E}$$

or

$$\sigma_t = \sigma_0 e^{-\frac{t}{\eta / E}} \qquad (1.5a)$$

where σ_0 and σ_t are the initial and current stress, respectively.

Expression (1.5*a*) dictates an exponential decrease in stress due to the elastic response of the spring, which is inhibited by the displacement of the piston in a viscous medium.

The ratio η/E in the exponent of the expression (1.5*a*) is a constant that controls the relaxation rate. This ratio has the dimension of time, and is called the relaxation time of the model τ. Taking this factor into account, expression (1.5*a*) is written in the form

$$\sigma_t = \sigma_0 e^{-\frac{t}{\tau}}.$$ (1.5b)

From the physical point of view, $\tau = \eta/E$ reflects the relationship between the viscous and elastic response of the system. The greater the viscosity of the fluid and the lower the elastic modulus of the spring, the greater the value of τ.

In the creep or strain ***relaxation*** experiments, the system at instant $t = 0$ is immediately loaded with a stress σ that is kept constant throughout the experiment. Under these conditions, $\sigma = $ const, $d\sigma/dt = 0$, and expression (1.4*a*) takes the form

$$\frac{d\varepsilon}{dt} = \frac{\sigma}{\eta} \text{ or } \frac{d\varepsilon}{dt} = \frac{\sigma}{E\tau}.$$

After integration we obtain

$$\varepsilon_t = \varepsilon_0 + \frac{\sigma}{\eta} t$$ (1.6a)

or

$$\varepsilon_t = \varepsilon_0 + \frac{\sigma}{E\tau} t.$$ (1.6b)

Expressions (1.6*a*) and (1.6*b*) indicate that under a constant load the deformation of the Maxwell model develops linearly, and the strain rate $\dfrac{d\varepsilon}{dt}$ is directly proportional to the applied stress σ and is inversely proportional to the fluid viscosity η, the elasticity modulus of the spring E and the relaxation time τ. This behaviour is shown in Fig. 1.1*b*.

Under the conditions of the experiment, at the initial instant of time, the model instantly responds to the action by elastic extension of the spring by an amount $\varepsilon_0 = \sigma/E$. Further deformation is caused

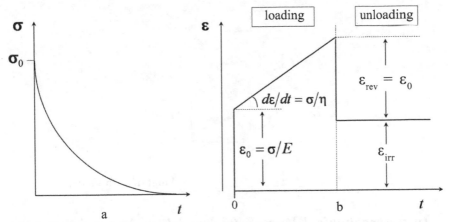

Fig. 1.1. The typical behaviour of the Maxwell model in the experiment on stress relaxation (*a*) and creep (*b*).

by the motion of the piston in a viscous fluid with a velocity equal to σ/η (Fig. 1.1*b*).

When unloading, i.e. when the stress is zeroed, the model instantaneously relaxes by compressing the spring, and the magnitude of the reversible strain ε_{rev} exactly corresponds to the elastic deformation $\varepsilon_{rev} = \varepsilon_0$.

Residual strain ε_{res} or, which is the same, the irreversible part of the total strain of the model ε_{irr} model is determined by the displacement of the piston in a viscous fluid.

Consequently, *the Maxwell model predicts the coexistence of reversible and irreversible deformation components*, and it is obvious that the total deformation is their sum:

$$\varepsilon = \varepsilon_{rev} + \varepsilon_{irr}.$$

Note that the ratio $\varepsilon_{rev}/\varepsilon_{irr}$ decreases with increasing loading time t: only elastic, reversible deformation is observed for $t \to 0$, and for sufficiently large t values the contribution of the irreversible component prevails.

The observed behaviour should be related to the ratio of the loading time t and the relaxation time of the model $\tau = \eta/E$, which can be varied by changing the elastic modulus of the spring and the viscosity of the fluid. For $\tau \ll t$ (a rigid spring and a liquid with a small viscosity), the model is characterized by irreversible strains. The reverse case (a soft spring and a liquid with high viscosity) is associated with the preferential elastic response of the model.

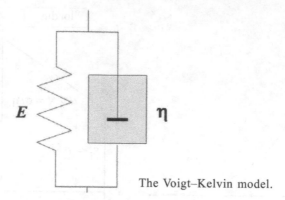

The Voigt–Kelvin model.

In the mechanical Voigt–Kelvin model, the same elements are connected in parallel, and the total deformation of the model as it stretches is equal to the elastic and viscous components:

$\varepsilon = \varepsilon_{el} = \varepsilon_\eta$, and the total stress is the sum of the stress σ_{el} necessary for the deformation of the spring and the stress σ_η required to move the piston in a viscous liquid:

$$\sigma = \sigma_{el} + \sigma_\eta.$$

Taking into account the laws of Hooke and Newton, we obtain

$$\sigma = E\varepsilon + \eta \frac{d\varepsilon}{dt} \tag{1.7a}$$

or

$$\frac{d\varepsilon}{dt} = \frac{\sigma - E\varepsilon}{\eta}. \tag{1.7b}$$

Under the deformation conditions with the action of a constant stress σ = const (creep), the integration of expression (1.7b) gives the following expression for the time dependence of the developing deformation:

$$\varepsilon_t = \frac{\sigma}{E}\left[1 - e^{-\frac{E}{\eta}t}\right] = \frac{\sigma}{E}\left[1 - e^{-\frac{t}{\tau}}\right]. \tag{1.8}$$

For this deformation behaviour, shown in Fig. 1.2 ('loading' curve), the strain rate decreases with time.

When the model is unloaded, when the stress becomes zero, expression (1.7b) is written as

$$\frac{d\varepsilon}{dt} = -\frac{E}{\eta}\varepsilon = -\frac{\varepsilon}{\tau}$$

After integration, we obtain the following expression for the relaxation of deformation:

$$\varepsilon_t = \varepsilon_0 e^{-\frac{E}{\eta}t} = \varepsilon_0 e^{-\frac{t}{\tau}} \tag{1.9}$$

Thus, *the Voigt–Kelvin model predicts the flow in time, i.e. the relaxation nature of reversible deformation* in contrast to the Maxwell model, within which reversible deformation develops and relaxes instantaneously.

We emphasize that the Voigt–Kelvin model does not describe the stress relaxation, since when the given deformation is kept constant ($d\varepsilon/dt = 0$) expression (1.7a) corresponds to Hooke's law, and only a linear elastic response is observed.

More complex mechanical models are different combinations of elastic and viscous elements. One of the most successful variants of combined models is shown below.

The general deformation of this model is the sum of:

- elastic (Hookean) strain ε_{el}, which is specified by a spring with a modulus of elasticity E_1;
- delayed reversible strain ε_{VK} due to expansion of the Voigt–Kelvin element with the elastic modulus E_2 and viscosity η_2;

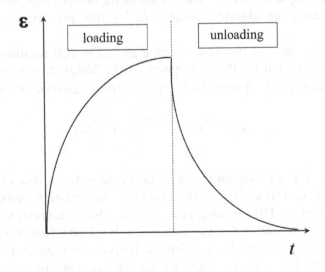

Fig. 1.2. Typical behaviour of the Voigt–Kelvin model under creep conditions.

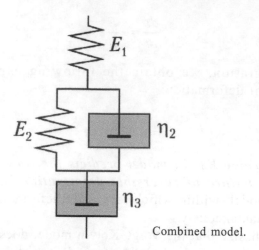

Combined model.

- the viscous component ε_η caused by the displacement of the piston in a liquid with viscosity η_3:

$$\varepsilon = \varepsilon_{el} + \varepsilon_{VK} + \varepsilon_\eta. \tag{1.10}$$

We obtain for the time dependence of the deformation:

$$\varepsilon_t = \frac{\sigma}{E_1} + \frac{\sigma}{E_2}\left[1 - e^{-\frac{t}{\tau}}\right] + \sigma\frac{t}{\eta_3}. \tag{1.11}$$

In the creep test ($\sigma = $ const), the loading of the combined model at time t_1 is accompanied by its instantaneous deformation by stretching the spring with the elastic modulus E_1 by the quantity $\varepsilon_{el} = \sigma/E_1$ (Fig. 1.3).

With an increase in the loading time to t_2, the deformation of the model is determined by the expansion of the Voigt–Kelvin element and the motion of the piston in a liquid with a viscosity of η_3:

$$\varepsilon_{VK} + \varepsilon_\eta = \frac{\sigma}{E_2}\left[1 - e^{-\frac{(t_2-t_1)}{\tau}}\right] + \sigma\frac{t_2-t_1}{\eta_3}.$$

When the model is unloaded at the instant t_2 ($\sigma \rightarrow 0$), an instantaneous elastic return or relaxation of the common deformation component ε_{el} (expression (1.10)) is observed due to the contraction of the spring with the modulus E_1. The delayed relaxation of the reversible deformation ε_{VK}, given by the Voigt–Kelvin element, occurs in the time period $t_3 - t_2$. In other words, the reversible part of the deformation of the model is $\varepsilon_{rev} = \varepsilon_{el} + \varepsilon_{VK}$.

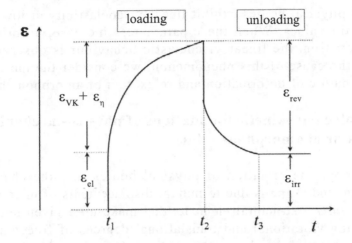

Fig. 1.3. Typical behaviour of the combined model under creep conditions,

It is obvious that the residual strain ε_{res} or the irreversible part of the total strain ε_{irr} is related to the displacement of the piston in a viscous element with viscosity η_3:

$$\varepsilon_{res} = \varepsilon_{irr} = \varepsilon_{\eta_3}.$$

Summarizing the above, we note that the considered mechanical models describe the following aspects of the viscoelastic behaviour of physical bodies:

1. Coexistence of a reversible and irreversible component of deformation;
2. The effect of loading time on their ratio, i.e. time-dependent relaxation character of the deformation;
3. The deformation, development and release (relaxation) of which occurs instantaneously (1) and delayed reversible deformation, developing and relaxing in time (2);
4. Stress and strain relaxation.

These aspects adequately reflect and predict the flow in time, i.e. the relaxation character of the deformation of the system, and also allow us to introduce a quantitative process parameter, namely, the 'relaxation time'. In this case, the activation characteristics and relaxation times of the models do not depend on external factors (primarily stresses and strains) and do not change during the test. This behaviour is called 'linear viscoelasticity', and the corresponding models are referred to as 'linear mechanical models'.

Real physical bodies exhibit linear viscoelasticity at low stress and strain values. When these parameters increase, a noticeable deviation from the linear viscoelastic behaviour is observed. To clarify the cause of this phenomenon, we consider the molecular-kinetic nature of deformation and relaxation of amorphous bodies.

1.2. Molecular–kinetic foundations of physico-mechanical behaviour of amorphous solids

Macroscopic deformation of physical bodies, i.e. the change in their size and shape is due to mutual displacements of microscopic *kinetic units* – structural elements that make up a given body and possessing vibrational and translational degrees of freedom. It is obvious that a molecule is a kinetic unit for low-molecular-weight amorphous compounds.

At a given temperature T in a solid, the molecules are capable of performing only oscillations near the equilibrium position. The amplitude of these oscillations is determined by the thermal energy kT. With increasing temperature, the amplitude of the oscillations naturally increases, and under the condition $kT \geq E_a$ it becomes possible to transfer molecules from one potential well to another, with the activation barrier E_a overcome. Such a transition is an elementary act of viscous flow under the action of its own weight.

The boundary temperature, starting from which the flow of a low-molecular-weight amorphous body manifests itself, is denoted as the flow temperature T_f (the temperature of the transition from the solid state to the liquid state) or as the glass transition temperature T_g (the transition temperature from the vitreous to the viscous flow state) (Fig. 1.4). For low-molecular-weight amorphous bodies these temperatures coincide, i.e. $T_g = T_f$.

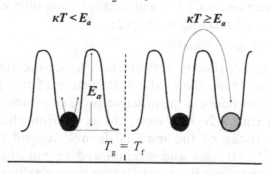

Fig. 1.4. Molecular flow mechanism.

The transition of a kinetic unit from one potential well to another (an elementary act of flow) does not occur instantaneously, but requires a certain time τ. In connection with this, the manifestation of the flow process is determined by the ratio of the observation time t and the time τ of translational motion of the kinetic unit.

If $t \gg \tau$, a pronounced flow develops during the observation time, and we perceive the given body as a fluid that retains its volume, but not its shape. Otherwise, when $t \ll \tau$, the time for elementary acts of transition of kinetic units from one position to another is not enough, and we perceive the given object as a solid body that preserves both volume and shape.

Consider the deformation of a solid, vitreous body under the action of applied stress at a temperature much lower than the glass transition temperature or flow temperature. In this temperature range only oscillations near the equilibrium position are thermally activated for kinetic units. In the conditions of external loading, the solid body is under the action of two fields – thermal kT and mechanical $\gamma\sigma$, where γ is a coefficient having the dimensionality of the volume.

If the sum of the thermal and mechanical energies is less than the value of the activation barrier ($kT + \gamma\sigma < E_a$), only a mutual displacement of the kinetic units relative to the equilibrium position is observed, which nevertheless leads to the appearance of a certain deformation $\varepsilon = \dfrac{l - l_0}{l_0}$ (Fig. 1.5).

At the same time, mechanical work is spent on overcoming the forces of intermolecular interaction. When unloading, the kinetic units return to their original position, which causes the reversible character of deformation.

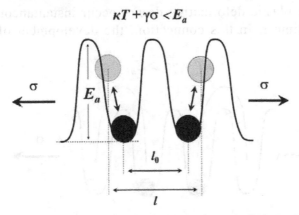

Fig. 1.5. Molecular mechanism of elastic deformation.

Such a deformation behaviour is elastic and obeys Hooke's law.

It is obvious that the elastic strains are of an energetic nature, since they are determined by the energy of the intermolecular interaction, and the modulus of elasticity is the measure of the elastic resistance of the body to external loads. As the temperature increases, the amplitude of the vibrations of the kinetic units increases, the intermolecular interaction weakens, and the elastic modulus of the body decreases, which dictates the experimentally observed inverse proportionality of $E \sim 1/T$.

If the sum of the thermal and mechanical energies is greater than the value of the activation barrier ($kT + \gamma\sigma > E_a$), then mutual displacements of the kinetic units due to their transitions to adjacent potential wells are possible. This is the deformation of the flow, which for solid bodies is called plastic deformation of yielding. Plastic deformation is irreversible, since during unloading the kinetic units do not return to the initial positions (Fig. 1.6).

As the temperature rises, plastic deformation occurs at a lower stress. These molecular–kinetic regularities were formalized by the American chemical physicist G. Eyring (for the flow of viscous fluids) [8, 9] and by the Soviet physicist A.P. Aleksandrov (for the plastic deformation of polymers) [10] in the form of expression

$$\dot{\varepsilon} = \dot{\varepsilon}_0 \exp\left[\frac{E_a - \gamma\sigma}{kT}\right].$$

where $\dot{\varepsilon} = \dfrac{d\varepsilon}{dt}$ is the flow velocity or strain rate; $\dot{\varepsilon}_0$ is the pre-exponent.

Transitions of kinetic units from one potential well to another that determine plastic deformation do not occur instantaneously, but for a certain time τ. In this connection, the development of plastic

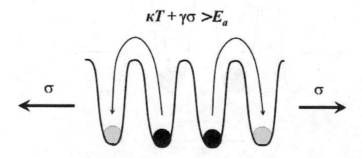

Fig. 1.6. Molecular mechanism of plastic deformation.

$$\left(t \sim \frac{1}{V}\right) \ll \left(\tau \sim \frac{1}{T}\right)$$

Elasticity criterion

Plasticity criterion

$$\left(t \sim \frac{1}{V}\right) \gg \left(\tau \sim \frac{1}{T}\right)$$

V – strain rate
T – deformation temperature

deformation is determined by the ratio of the time τ and the loading tim t inversely proportional to the strain rate.

Even under the condition $kT + \gamma\sigma > E_a$, at high impact rates, when $t \ll \tau$, the plastic deformation does not have time to develop, and a purely elastic response of the material to the external action is observed. Plastic deformation is realized only under the condition $t \gg \tau$ at the corresponding strain rate.

The relationship between the observation time or the time of mechanical action t and the time of translational motion of the kinetic unit τ is determined by the Deborah number introduced by the Austrian engineer M. Reiner: $De = \tau/t$ [11].

The material is considered perfectly elastic if $De \to \infty$ ($t \ll \tau$). For an ideal fluid, $De \to 0$ ($t \gg \tau$), and for viscoelastic bodies (primarily for polymers) $De \sim 1$ ($t \sim \tau$).

Thus, for low-molecular bodies there are two basic types of strain – elastic and plastic. The nature of elastic reversible strains is associated with the vibrational displacements of the kinetic unit (molecule) near the equilibrium position. Flow strains or plastic strains develop due to translational movements of the kinetic unit from one potential well to another with the overcoming of the activation barrier.

In the framework of the relaxation approach, the criterion for the transition from elastic strains to flow (or plasticity) is the condition $t \sim \tau$. The time t is determined by the rate of deformation, with an

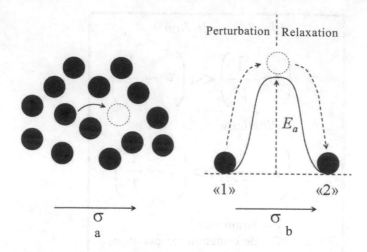

Fig. 1.7. Schematic representation of the elementary act of flow as a transition of the kinetic unit to the adjacent cavity of the free volume (*a*) with the overcoming of the activation barrier (*b*).

increase in which t decreases. Time τ depends on the deformation temperature and decreases as it grows.

In the general case, the viscoelastic behaviour of amorphous bodies is a combination of elasticity and flow. The flow (plasticity) begins to prevail when, at a fixed temperature, the rate of deformation is reduced or, at a fixed rate, the deformation temperature is increased. Conversely, a viscoelastic body exhibits elastic properties under reverse changes in the temperature–velocity regimes of deformation.

In other words, an increase in the deformation temperature is equivalent to a decrease in the strain rate. This conclusion underlies the principle of temperature–time superposition, which is considered in detail in Section 1.4.

The above molecular–kinetic considerations do not take into account the fact that the transition of a kinetic unit to an adjacent potential well requires the presence of a cavity or a pore of a free volume (Fig. 1.7*a*).

Overcoming the activation barrier caused by the reactions of neighboring kinetic units includes

- 'perturbation' of the system, i.e. its derivation from the initial equilibrium state (potential well «1»);
- Subsequent discharge or relaxation of the 'perturbation' due to the transition of the particle to another equilibrium state (potential well «2»).

Consequently, the deformation of the flow or plastic deformation associated with the translational motion of the kinetic units is a continuous set of elementary acts of 'perturbation–relaxation».

1.3. Viscoelasticity and relaxation properties of amorphous polymers

Within the framework of the molecular–kinetic analysis of the physical–mechanical behaviour of a material, the first task is to determine the kinetic unit. As noted in Section 1.2 for low-molecular-weight amorphous bodies, the kinetic unit is a molecule. For polymers, the situation seems more complicated.

It is obvious that macromolecules (macromolecular coils), capable of moving relative to each other, can serve as kinetic units.

The polymer macromolecule is characterized by a chain structure, i.e. a sequence of chemical bonds around which internal rotation is possible. In the individual macromolecule, kinetic segments can be distinguished – the minimal portion of a chain capable of quasi-independent translational movements, activated by temperature or external stress.

Thus, unlike low-molecular-weight bodies in polymers, there are two types of kinetic units – the kinetic segment and the macromolecular coil.

Experimentally, the difference in the nature of the kinetic units of a low-molecular-weight body and a polymer is clearly recorded when comparing their thermomechanical curves, i.e. temperature dependences of deformation ε, which develops in the sample under the action of a given stress (Fig. 1.8)[1].

1)The principles of thermomechanical analysis are discussed in detail in Appendix 1.

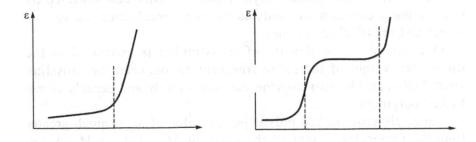

Fig. 1.8. Typical thermomechanical curves of a low-molecular-weight amorphous body (*a*) and an amorphous polymer (*b*).

For low-molecular-weight amorphous bodies (Fig. 1.8a), at a glass transition temperature T_g equal to the flow temperature T_f, the mobility of a single kinetic unit, the molecule, is 'unfrozen'. As a result, the material's deformability increases dramatically.

For an amorphous polymer (Fig. 1.8b), these two temperatures do not coincide, and a purely polymeric physical, high elastic or rubbery state is realized between them. The nature of the observed effect is due to the presence of two types of kinetic units in the polymer. At the glass transition temperature, only the mobility of the kinetic segments is 'unfrozen', and the 'thawing' of the mobility of macromolecular coils occurs only at a flow temperature.

We note that the glass transition temperature T_g of a polymer is determined by its chemical structure and does not depend on the molecular weight, i.e. is a characteristic parameter. The flow temperature T_f of the polymer increases as the molecular weight increases. As a consequence, as the molecular weight increases, an expansion of the temperature range of high elasticity is observed.

The reason for this behaviour is due to the fact that the critical length of the chain, from which a highly elastic state appears, is comparable to the length of the segment. Further growth of the length of the macromolecule and, consequently, of the molecular mass leads only to an increase in the number of segments and does not affect the activation parameters of this kinetic unit, and, therefore, the value of T_g. In this case, the increase in molecular mass is accompanied by a noticeable increase in the size of the macromolecular coil. As a result, the activation barrier that controls the translational motion of these kinetic units increases, the development of flow deformation requires ever higher temperatures, and T_f naturally increases.

The determining role of this or that kinetic unit (segment or macromolecular coil) in the formation of the physico-mechanical behaviour of an amorphous polymer can be easily demonstrated by performing a comparative analysis of thermomechanical curves for linear and crosslinked samples.

Quantitatively, the density of crosslinking is estimated as the molecular weight of the chain fragment M_c between the crosslink sites. Let's consider two extreme variants – rarely and densely cross-linked polymers.

For rately cross-linked polymers, the value of M_c is much greater than the molecular weight of the segment M_{segm} ($M_c \gg M_{segm}$), i.e. the chain fragment between the cross-link nodes includes a plurality of segments. In this case, cross-linking has no effect on segmental

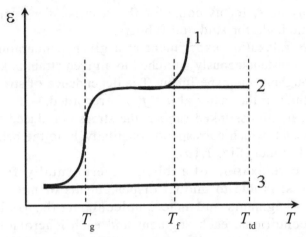

Fig. 1.9. Typical thermomechanical curves of linear (1), rately cross-linked ($M_c \gg M_{segm}$) (2) and a densely cross-linked amorphous polymer ($M_c < M_{segm}$) (3).

mobility, and macromolecular coils, bound by chemical 'bridges' into a single three-dimensional grid, lose their mobility. In other words, in such materials (primarily vulcanized rubbers) there is only one kinetic unit – the segment. This naturally leads to the degeneration of the viscous flow state and the expansion of the temperature range of high elasticity up to the thermal decomposition temperature of the polymer T_{td} (Fig. 1.9, curve 2).

For the densely cross-linked polymers, $M_c < M_{segm}$, resulting in the immobilization of both segments and macromolecular coils. As a result, densely cross-linked resins exist only in the vitreous state (Fig. 1.9, curve 3), and their response to mechanical action is determined only by elastic deformation.

The presence in the polymer of two types of kinetic units controls the following mechanisms of flow deformation are possible:

1. Deformation due to the translational motion of macromolecular coils relative to each other, i.e. due to the displacement of their centres of mass;

2. Specific deformation due to the unfolding of the coils, controlled only by translational movements of segments (segmental mobility) without the displacement of the mass centres of the macromolecular coils;

3. Deformation, which includes a combination of the two above cases.

For a more detailed analysis of the deformation behaviour of amorphous polymers, let us consider the results of a number of experimental methods for studying rubbers.

In the stress relaxation experiment at a given temperature, the rubber sample is instantaneously stretched to a given strain ε, keeping it constant throughout the experiment. The dependence of stress σ as a function of time in the material $\sigma = f(t)$ is recorded.

For a linear, non-crosslinked rubber, the stress is reduced to zero (Fig. 1.10, curve *1*), which corresponds qualitatively to the behaviour of the Maxwell model (Fig. 1.1*a*).

Macroscopic relaxation of rubber, experimentally fixed by decreasing stress, is due to mutual displacements of microscopic kinetic units – segments and macromolecular coils. Under the experimental conditions, each segment and each macromolecular coil tend to 'relax', i.e. go to the most equilibrium state. Elementary transitions of segments and coils occur in a viscous medium of their own and require a certain time, which should be regarded as the relaxation time of this segment and the relaxation time of this macromolecular coil – τ_s and τ_{mc}, respectively.

In a real material, each segment and each macromolecular coil have different environments which determines the presence of a spectrum of relaxation times of segments and the spectrum of relaxation times of macromolecular coils. In this regard, for linear rubber, the experimental dependence of stress on time (Fig.

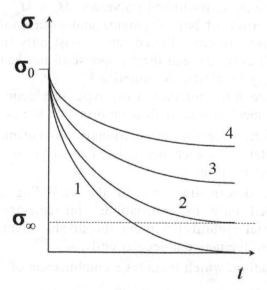

Fig. 1.10. Typical stress relaxation curves for linear (1) and crosslinked rubbers (2–4). $M_c(2) > M_c(3) > M_c(4)$.

1.10, curve *1*) is described not by a monoexponential of the type of expression (1.5*b*), but by a continuous set of exponentials. To simulate the spectrum of relaxation times, it is necessary to connect in parallel a set of Maxwell models with different values of the elasticity modulus of the spring E and the viscosity of the liquid η.

Rare crosslinking or vulcanization of rubber completely excludes mutual displacements of macromolecular coils. For such materials, both deformation and relaxation are determined only by segmental mobility. As a result, complete relaxation does not occur, and the stress decreases to a certain quasi-equilibrium value of stress σ_∞ (Fig. 1.10, curve *2*). With an increase in the density of the cross-linking or a decrease in the length of the chain M_c between the nodes of the chemical network, the value of σ_∞ increases (Fig. 1.10, curves *2–4*).

In a creep experiment at a fixed temperature, a rubber sample is instantaneously loaded to a predetermined value of stress σ, keeping it constant throughout the experiment. We record the dependence of the strain ε in the polymer as a function of time: $\varepsilon = f(t)$. A typical creep curve for a linear non-crosslinked rubber is shown in Fig. 1.11 (curve *1*).

At the initial part of this curve, the deformation versus time dependence is close to linear. Further, a deviation from linearity is observed with a subsequent exit to the next linear section, which is called the region of steady flow.

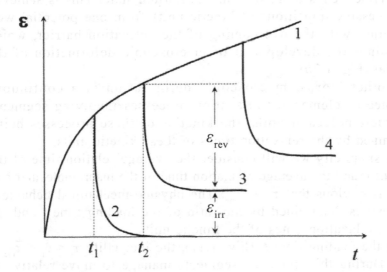

Fig. 1.11. A typical creep curve for linear rubber (1) and unloading curves for deformed samples (2–4). Explanations in the text.

We shall complicate the experiment as follows. A sample is deformed for a time t_1, then the stress is removed. The deformed rubber tends to return to the initial state, and the given deformation gradually decreases (Fig. 1.11, curve 2). In this case, a complete and reversible strain recovery to a zero value is observed, but this process requires a certain time.

The increase in the time of mechanical action to the value t_2 radically changes the picture observed after unloading (Fig. 1.11, curve 3). A complete deformation recovery does not occur, and a given deformation includes two components: reversible (ε_{rev}) and irreversible (ε_{irr}). A further increase in the loading and deformation time is accompanied by an increase in the contribution of the irreversible component to the total deformation (Fig. 1.11, curve 4).

Thus, in the general case, under the conditions of the creep test, the total deformation can be represented as the sum of a reversible and an irreversible component:

$$\varepsilon = \varepsilon_{rev} + \varepsilon_{irr},$$

and the appearance and growth of an irreversible component is determined by the loading time. The molecular–kinetic picture of the observed behaviour reduces to the following.

At each moment of time, the constantly acting stress 'removes' the rubber sample from the equilibrium state, and at each instant the polymer tends to go into an equilibrium state. This is achieved by successive transitions of kinetic units from one potential well to another with the overcoming of the activation barrier, which determines the development of macroscopic deformation of the material (Fig. 1.7b).

In other words, macroscopic deformation is a continuous sequence of elementary relaxation processes involving segments and macromolecular coils, the kinetics of these processes being determined by the relaxation times of these kinetic units.

For simplicity we will consider the average relation time of the segment $\bar{\tau}_s$ and the average relaxation time of the macromolecular coil $\bar{\tau}_{mc}$. It is obvious that $\bar{\tau}_s \ll \bar{\tau}_{mc}$. The physico-mechanical behaviour of rubber is determined by the ratio of the loading time t and the average relaxation times of the kinetic units.

For the loading time t_1 (Fig. 1.11), the inequality $\bar{\tau}_s < t_1 < \bar{\tau}_{mc}$ is valid. During this time, the segments manage to move relative to each other, determining the development of reversible high-elastic deformation. At the same time, the coils themselves remain in place, as time for their mutual movement is not enough. After removing the load, the deformed coils naturally take their original dimensions, causing the observed reversibility of deformation (Fig. 1.11, curve *2*).

Elementary processes of reversible 'return' of deformed coils to their original dimensions and shape are determined by segmental mobility. The displacements of the segments take place in a viscous environment of their own, which causes a delayed nature of the relaxation of the high elastic deformation. This behaviour is satisfactorily described by the Voigt–Kelvin model (Fig. 1.2), in which a spring with a module E controls the reversibility of deformation, and the motion of the piston in a fluid with viscosity η has a retarding effect.

The loading time t_2 is greater than $\bar{\tau}_s$ and $\bar{\tau}_{mc}$. During this time macromolecular coils have time to turn around and move relative to each other. After removing the load, as in the previous case, a reversible (high elastic) deformation component is observed. However, the mutual displacement of macromolecular coils is irreversible, as a result of which an irreversible macroscopic deformation component appears, whose contribution increases with increasing loading time (Fig. 1.11, curves *3* and *4*).

Summing up, we note that an analysis of the experimental results of stress and strain relaxation together with model representations allows us to formulate the following aspects of the viscoelastic behaviour of rubbers:

- delayed development and relaxation of reversible (high elastic) deformation;
- coexistence of reversible (high elastic) deformation and irreversible deformation of the flow.

In this case, the mechanism of high elasticity is due to deformation of macromolecular coils due to segmental mobility, and the mechanism of viscous flow is due to mutual movements of coils.

For vulcanized (cross-linked) rubbers, mutual displacements of macromolecular coils are excluded. As a result, both deformation and relaxation of such materials is determined only by segmental mobility. In this connection, on the creep curves, the region of steady flow is not observed, and the deformation is completely reversible at any loading times (Fig. 1.12).

For further investigations of the viscoelastic behaviour of rubbers, let us consider the results of their cyclic deformation in the 'stretching–shrinking' mode.

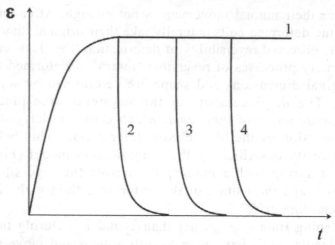

Fig. 1.12. A typical creep curve for cross-linked rubber (1) and unloading curves for deformed samples (2 to 4).

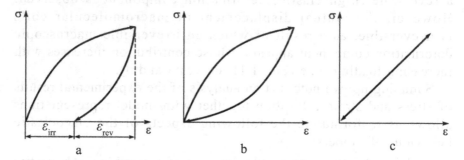

Fig. 1.13. The 'stretch–shrink' cycles for linear rubber (*a*), cross-linked rubber (*b*) and ideally elastic body (*c*).

In cyclic tests, the rubber sample is stretched at a constant rate to a certain amount of strain or stress, after which the unloading (contraction) is carried out at the same speed until the zero value of the stress is reached. The interval of strains corresponding to the linear viscoelasticity of the material is chosen, where structural changes caused by force are not observed and, consequently, the activation parameters of the deformation processes do not change.

Figure 1.13*a* shows a typical 'stretch–shrink' diagram for a linear non-crosslinked rubber. When unloading to the zero stress value, the sample is characterized by a significant amount of residual irreversible deformation (ε_{irr}), the nature of which is related to the irreversible viscous flow of macromolecular coils. The high elastic reversible part of deformation (ε_{rev}) is restored at unloading due to the segmental mobility of macromolecules. Thus, the total deformation of the non-crosslinked rubber can be represented as the sum of the above components:

$$\varepsilon = \varepsilon_{rev} + \varepsilon_{irr}.$$

In this case, the tension and contraction curves do not coincide, which leads to the appearance of a hysteresis loop.

As already mentioned above, vulcanization (rare cross-linking) of rubber completely excludes mutual translational movements of macromolecular coils, i.e. viscous flow. Deformation develops only by the mechanism of high elasticity due to segmental mobility, and as a result, unloading of the crosslinked polymer is accompanied by zeroing of the given deformation and complete restoration of the initial size (Fig. 1.13*b*). However, this type of deformation is also characterized by a noticeable hysteresis.

The methodology of the experiment on cyclic deformation and interpretation of the results obtained is described in detail in Appendix 1. Here, for the sake of maintaining the integrity of the presentation, we confine ourselves to the following general remarks.

So, in the cyclic tests

- *the area under the extension curve S* is the deformation work per unit volume of the sample.

$$S = \int_0^\varepsilon \sigma d\varepsilon, \text{ where } \sigma d\varepsilon = \frac{f}{S_{c.s.}} \frac{dl}{l_0} = \frac{fdl}{V} = \frac{A_d}{V},$$

where f is the force, $S_{c.s.}$ the cross-sectional area of the sample, V is the initial volume of the sample, and A_d is the deformation work.

- *the area under the contraction curve characterizes* the part of the spent work A_{el}, which the physical body reversibly (elastically) returns when unloading.
- *the area of the hysteresis loop characterizes* that part of the work expended which is irreversibly 'lost' during a cyclic test, A_{loop}.
- the *observed mechanical* losses are quantitatively estimated by the coefficient of mechanical losses $\chi = \dfrac{A_{loop}}{A_d}$.

Let us compare the viscoelastic behaviour of rubbers (Figs. 1.13*a* and *b*) with the behaviour of an ideally elastic, Hookean body (Fig. 1.13*c*). For the latter, both the geometric reversibility of deformation (complete restoration of the original size and shape after unloading) and the thermodynamic reversibility are characteristic (the forward and reverse processes pass through the same intermediate states). In other words, all the work expended for deformation is completely returned during 'shrinking', and mechanical losses are absent ($\chi = 0$). We note that when the ideal fluid is deformed (viscous flow), the entire work expended for deformation is completely 'lost' ($\chi = 1$).

For a linear rubber (Fig. 1.13a), neither geometrical nor thermodynamic reversibility is observed. Vulcanized rubber (Fig. 1.13b) demonstrates the geometric, but not thermodynamic, reversibility of deformation.

The nature of hysteresis phenomena or mechanical losses observed for polymers under cyclic action is associated with translational movements of macromolecular coils and segments. As a result, internal friction occurs in the sample, and mechanical work is partially dissipated as heat. In contrast, the elastic deformation of the Hookean body is determined by the deviation of the kinetic units from the equilibrium position without translational displacements. At the same time, internal friction does not occur, mechanical losses are not observed and there is no hysteresis.

A viscoelastic body that combines both elastic deformations and deformations of a viscous flow is characterized by a coefficient of mechanical losses lying in the range $0 < \chi < 1$. In this case, the area under the contraction curve corresponds to the elastic component of the deformation A_{el}, and the area of the hysteresis loop is the component associated with the viscous deformation component A_{loop}. The work expended on deformation is represented as the sum of these two components:

$$A_d = A_{el} + A_{loop}.$$

The coexistence of these components is a distinctive feature of the viscoelastic body.

The presence of hysteresis, i.e. the mechanical losses, is a sign of the process's non-equilibrium, and the area of the hysteresis loop and the value of the coefficient of the mechanical losses serve as a quantitative measure of the non-equilibrium. In equilibrium processes, for example, when a perfectly elastic body deforms (Fig. 1.13c), at each instant of time the system reaches equilibrium and there is no hysteresis.

Thus, the signs of viscoelasticity are:

1. coexistence of reversible and irreversible components of

deformation $\varepsilon = \varepsilon_{rev} + \varepsilon_{irr}$;

2. coexistence of the elastic and viscous components of the mechanical work $A_d = A_{el} + A_{loop}$.

Note that the last characteristic is also characteristic for the case of complete geometric reversibility of deformation, when $\varepsilon = \varepsilon_{rev}$ (Fig. 1.13*b*).

1.4. The principle of temperature–time superposition

The principle is based on the Deborah criterion or number $De = \tau/t$ (see Section 1.2), according to which the mechanical response of a material is determined by the ratio of two time characteristics:

1. Time t or frequency $\omega \sim \dfrac{1}{t}$ or speed of mechanical action $V \sim \dfrac{1}{t}$;
2. The time of the elementary deformation act τ, i.e. time transition of the kinetic unit responsible for the development of deformation, from one state to another. This time, interpreted as the relaxation time of the kinetic unit, is inversely proportional to temperature:

$$\tau = \tau_0 e^{\frac{E_a}{RT}},$$

where E_a is the activation energy of the process, and τ_0 is the pre-exponent.

Consequently, one and the same mechanical response of the material can be achieved

- either by changing the exposure time at a constant temperature, i.e. with a constant relaxation time;

- or by changing the relaxation time due to temperature changes at a constant exposure time.

The principle of temperature–time superposition is widely used for predicting the mechanical behaviour of a material at exposure frequencies that are inaccessible for direct study since modern dynamometers allow performing dynamic mechanical tests in a rather limited frequency interval ω, which is 4–5 orders of magnitude. Let us explain this procedure using the example of the frequency dependences of the elastic modulus of rubber obtained at different temperatures (Fig. 1.14).

On the frequency dependence obtained at the selected temperature (the reference temperature T_{ref}), for example, T_1 (Fig. 1.14), we select a certain frequency of the action (the reference frequency ω_{ref}), for example, ω_2. For a given pair of test parameters, the mechanical

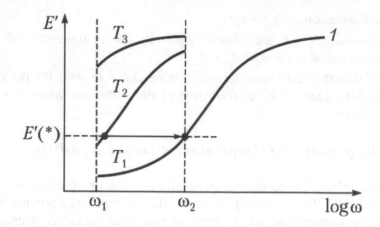

Fig. 1.14. The experimental frequency dependences of the elastic modulus E' in the frequency interval $\omega_1-\omega_2$ at temperatures $T_1 > T_2 > T_3$ and the mater curve (curve 1).

response of the material is characterized by a strictly defined value of the elastic modulus $E'(*)$. For another test temperature, for example, for T_2, a frequency ω is found at which the same value of the modulus of elasticity is observed.

The superposition of these two dependences is carried out by parallel transfer of the curve obtained at T_2 to the right by the value (log ω_{red} – log ω). The magnitude of the parallel transfer is determined by the factor of shift a_T:

$$\log a_T = \log \omega_{ref} - \log \omega = \log\left(\frac{\omega_{ref}}{\omega}\right).$$

The repetition of this procedure for the subsequent pair of curves obtained in our case at T_2 and T_3 leads to the construction of a master curve at the reference temperature (Fig. 1.14, curve *1*) in the frequency range not available for direct experimental studies.

To illustrate the practical possibilities of the temperature–time superposition principle, let us consider the following experimental results. Figure 1.15*a* shows the frequency dependences of the loss tangent tg δ (internal friction spectra) of poly(methyl methacrylate) obtained at different temperatures [12]. The interval of used frequencies (log $\omega = -1 \div 2$) was limited by the capabilities of the device.

The reference temperature was $T = 130°C$ (curve *3*, Fig. 1.15). In accordance with the procedure described above, the dependences

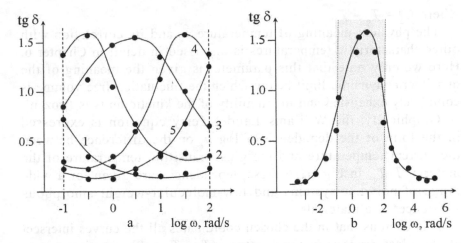

Fig. 1.15. Experimental frequency dependences of the loss tangent tg δ for poly(methyl methacrylate) obtained at temperatures of 110°C (1), 120°C (2), 130°C (3), 140°C (4) and 150°C (5), *a*) and the master at the reference temperature T_{red} = 130°C (*b*).

obtained at lower temperatures (curves *1* and *2*) were shifted to the right, i.e. towards higher frequencies, and the dependences obtained at higher temperatures (curves *4* and *5*) – to the left, i.e. in the direction of lower frequencies. This made it possible to construct a master curve (Fig. 1.15*b*), which describes the behaviour of the material at 130°C in the frequency range much wider than the experimentally allowed frequency interval (the darkened region in Fig. 1.15*b*).

In general, the shift factor depends on the reference temperature. A semi-empirical Williams–Landel–Ferry equation (WLF) is used to describe this dependence [13]:

$$\log a_T = \log\left(\frac{\tau}{\tau_{ref}}\right) = \log\left(\frac{\eta}{\eta_{ref}}\right) = \frac{-C_1(T - T_{ref})}{C_2 + (T - T_{ref})} = -C_1\frac{T - T_{ref}}{T - T_0},$$

where C_1 is a constant, and τ and η are the relaxation time and viscosity at the current temperature T, τ_{ref} and η_{ref} are the relaxation time and viscosity at T_{ref}, $C_2 = T_{red} - T_0$.

When the glass transition temperature T_g is selected as the reduction temperature, the WLF equation is written in the form:

$$\log a_T = \log\left(\frac{\tau}{\tau_g}\right) = \log\left(\frac{\eta}{\eta_g}\right) = \frac{-C_1(T - T_g)}{C_2^g + (T - T_g)} = -C_1\frac{T - T_g}{T - T_0},$$

where $C_2^g = T_g - T_0$.

The physical meaning of temperature T_0 and its correlation with other characteristic temperatures is discussed in detail in Chapter 6. Here we only note that this parameter is given the meaning of the quasithermodynamic limit below which the fluctuation free volume is completely exhausted and the mobility of the kinetic units is 'frozen'.

Graphically, the Williams–Landel–Ferry equation is expressed in the form of the depndence of log a_T on the difference between the current temperature and the glass transition temperature of the material $T-T_g$. In Fig. 1.16 these dependences are given for a wide range of polymer systems and low-molecular-weight amorphous substances and materials.

It is obvious that in the chosen coordinates all the curves intersect at the glass transition temperature $(T - T_g = 0)$, fanning out at temperatures above and below the glass transition temperature.

In recent decades, numerous attempts have been made to determine the universal form of the temperature dependence of the

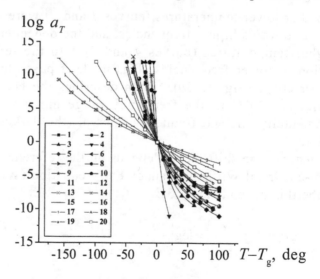

Fig. 1.16. Dependences of the function a_T on the difference between the current temperature and the glass transition temperature $T - T_g$ for a number of amorphous materials: polyisobutylene (1), poly(vinyl acetate) (2), poly(methyl acrylate) (3), polyhexene-1 (4), natural rubber (5), poly(butyl methacrylate) (6), poly(ethyl methacrylate) (7), poly(hexyl methacrylate) (8), poly(octyl methacrylate) (9), polyurethane 10), poly(butyl methacrylate) (50%) in diethyl phthalate (11), poly(butyl methacrylate) (60%) in diethyl phthalate (12), glycerol (13), selenium (14), Na_2O (36.2 mol%) – SiO_2 (15), Na_2O (19.0 mol%) – SiO_2 (16), Na–Ca–Si (17), K_2O (8.5 vol.%) –B_2O_3 (18), K_2O (8.5 vol.%) –B_2O_3(19), Na_2O (5 mol .%) – GeO_2(20).

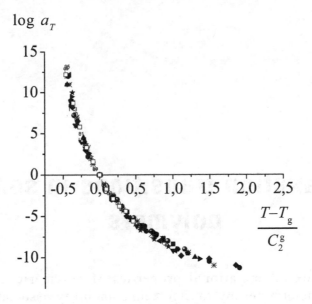

Fig. 1.17. The unified temperature dependence of the function a_T for amorphous materials. The notation is the same as in Fig. 1.16.

function a_T for various systems with the aim of creating a general picture of the viscoelastic behaviour of amorphous bodies. One of the variants of such unification [14, 15] involves the treatment of temperature dependences in the reduced dimensionless coordinates

$$\lg a_T = f\left[\frac{(T - T_g)}{C_2^c}\right] \quad \text{(Fig. 1.17)}.$$

The unified curve obtained unambiguously indicates the generality of the viscoelastic behaviour of amorphous glassy bodies, as well as the nature and mechanism of the vitrification of these substances and materials, regardless of their individual chemical structure, which is evidence of the universal character of the Williams–Landel–Ferry equation and significantly extends its predictive character.

2

Relaxation transitions in some polymers

The complex of operational properties of polymeric materials is largely determined by the molecular mobility associated with the manifestation of a wide range of torsional–vibrational and translational motions of kinetic units of various sizes.

The realization of the mobility of a particular kinetic unit in a given temperature–time test mode determines the course of a relaxation transition accompanied by a change in the response of the material to an external action. Obviously, understanding the nature and regularities of these relaxation transitions is essential for predicting the behaviour of a material under operating conditions and developing scientific and technological principles for creating materials with the required set of properties.

The range of experimental approaches and techniques for the study of relaxation phenomena is extremely diverse and includes thermomechanical analysis, differential scanning calorimetry, dielectric and acoustic measurements, radio thermoluminescence, nuclear magnetic resonance, various modifications of probe methods, etc. Already by the 1970s and 1980s, a huge amount of factual material was accumulated, reflecting the physical, mechanical and structural aspects of relaxation. The results of these studies for specific polymers are given in a large number of original articles and a number of monographs, for example, [16–21]. A detailed analysis of such an array of data is beyond the scope of this book, therefore, in this chapter we shall single out only the main features of the relaxation behaviour of polymers.

First of all, we note that the relaxation transitions can be divided into two groups.

The first group includes transitions that can be described in terms of the Arrhenius equation

$$v = v_0 \exp\left(-E_a \middle/ RT\right) \tag{2.1a}$$

or

$$\tau = \tau_0 \exp\left(E_a \middle/ RT\right), \tag{2.1b}$$

where E_a is the activation energy, v is the frequency of the transition of the kinetic unit from one position to another, τ is the time of the given transition or the relaxation time ($v = 1/\tau$), v_0 and τ_0 are the pre-exponentials.

The sign of the classification of the relaxation transition to the Arrhenius type is the linearity of the dependences $\lg v = f(1/T)$ and $\lg \tau = f(1/T)$, which for $1/T \to 0$ is extrapolated with the accuracy to 1–2 orders of magnitude to $v_0 \approx 10^{13}$ Hz and $\tau_0 \approx 10^{-13}$ s.

The Arrhenius type includes kinetically simple, small-scale, non-cooperative transitions.

The second group is formed by transitions whose behaviour does not obey the Arrhenius equation, and the experimental dependences $\lg v = f(1/T)$ and $\lg \tau = f(1/T)$ deviate significantly from linearity. In polymers, the nature of such transitions is interpreted from the standpoint of a cooperative, i.e. correlated, movement of segments and supersegmental structures.

The assignment of a specific transition to a particular type is discussed below when discussing the relaxation spectra of different classes of polymer materials.

2.1. Amorphous polymers

Amorphous polymers, i.e. polymers characterized by short-range order in the arrangement of atoms and atomic groups, exist in three physical states – vitreous, high elastic and viscous flow, separated by the glass transition temperature T_g and flow temperature T_f. The specific behaviour of the polymer in these states, as well as the transitions from one state to another, are determined by the ratio of the relaxation times of segments and macromolecular coils, on the one hand, and the exposure time on the other (see Section 1.3).

In connection with this, these physical states are interpreted as relaxation states, giving transitions from state to state the meaning of relaxation transitions.

The variety of relaxation spectra of amorphous polymers (see, for example, [21]) is dictated by the chemical structure of a particular sample, its prehistory, processing conditions, preparation, etc. Nevertheless, an analysis of the results of numerous original studies makes it possible to single out general and characteristic patterns of their relaxation behaviour.

Figure 2.1 shows a typical spectrum of internal friction (the temperature dependence of the loss tangent tg δ) for an amorphous polymer.

The main relaxation α-transition is related to the glass transition of the polymer and refers to cooperative transitions, to which the Arrhenius equation is not applicable. At temperatures below $T_\alpha = T_g$ the amorphous polymer is in a glassy state characterized by a set of secondary Arrhenius, relaxation β-, γ-, and δ-transitions.

Above T_α in the high elastic state of an amorphous polymer a 'liquid – liquid' transition is observed at a temperature T_{ll}, which, like the α-transition, can not be described in terms of the Arrhenius equation.

We note that many secondary μ-, π- and λ-transitions of the Arrhenius type [21] are observed in the temperature range $T > T_\alpha$.

The nature of the secondary relaxation γ- and δ-transitions in the glassy state ($T < T_\alpha$) is interpreted from the point of view of the 'thawing' of the mobility of the side groups or their fragments,

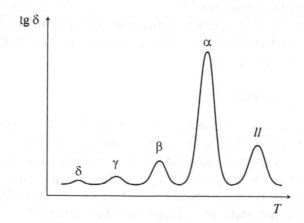

Fig. 2.1. Schematic representation of the internal friction spectrum for an amorphous polymer.

as well as the portions of the macrochain with a size smaller than the segment. In the rubbery state ($T > T_\alpha$), the fixed μ-, π- and λ-transitions are associated with the destruction of the nodes of the fluctuation structure.

Below, attention is concentrated on the description of the three main relaxation transitions (β-, α- and ll-), which have the most noticeable effect on the formation of the physicomechanical and operational properties of amorphous polymers.

2.1.1. Relaxation β-transition

The relaxation β-transition closest to the glass transition temperature (Fig. 2.1) is an Arrhenius-type transition. Depending on the chosen method and test modes (rate of change of temperature and frequency of mechanical loading), this transition is resolved in the form of either a single peak or a shoulder on the low-temperature wing of the α-peak.

The profiles of the relaxation spectrum and the intensity of the β-transition can vary greatly when passing from one polymer to another, and also depending on the prehistory of the samples. This is observed at frequencies of exposure not exceeding 10^6–10^7 Hz. With an increase in frequencies above 10^7–10^8 Hz, the α- and β-transitions merge into a single peak.

The nature of the β-transition is still widely discussed in the literature [3,4,16–22]. Various authors associate this transition with the motions of short (smaller than a segment) sections of the main chain, the mobility of the side groups and their fragments, the presence of impurities, etc. The proposed variants of the mechanism naturally raise the question of the inter- or intramolecular origin of this relaxation process.

It should be emphasized that in the 70s of the last century it was shown that the manifestation of β-relaxation is not specific for polymers and is also observed in low-molecular-weight vitrified liquids [23–25]. The observed processes were associated with molecular motions in structural regions with a reduced packing density. The subsequent development of these studies allowed us to conclude that the manifestation of the β-relaxation process as a precursor of the α-transition is a characteristic feature of the disordered bodies.

In [20,26–28], a correlation of the activation parameters of β-relaxation with the following characteristics is discussed:

- activation parameters of viscous flow;
- the magnitude of the Kuhn segment;
- barrier of internal rotation;
- cohesion energy.

It is well known that the dependence of the activation energy of the viscous flow for polymer homologous series on the degree of polymerization goes to saturation when the length of the rheological segment is reached [29, 30]. For a number of polymers, analogous dependences were also observed for the β-transition [26–28]. It is found that the length of the chain, starting from which the activation energy of the β-relaxation reaches a constant value, is comparable with the value of the statistical segment. Moreover, for flexible-chain polymers, the activation energy of a viscous melt flow with an accuracy of $\pm 10\%$ corresponds to the activation energy of the β-process. The set of correlations obtained indicates the segmental nature of the elementary act of a given relaxation transition.

A correlation was found between the activation energy of the relaxation β-transition E_a^β and the cohesion energy E_c for a wide range of low-molecular-weight vitrified liquids sharply differing in the nature of the intermolecular interaction:

$$E_a^\beta \approx (0.4 \pm 0.1) E_c. \tag{2.2}$$

Expression (2.2) is in good agreement with the well-known Eyring relationship, which determines the relationship between the activation energy of the viscous flow of simple liquids E_a^η and their cohesive energy:

$$E_a^\eta \approx (0.3 \pm 0.05) E_c. \tag{2.3}$$

Comparison of the expressions (2.2) and (2.3) indicates that the activation barrier for translational motion of molecules in a condensed medium is about one-third of the total energy of intermolecular interactions, and the molecular acts of the viscous flow of low-molecular-weight liquids and their solid-state β-relaxation have obvious similarities.

For the polymers, the following relationship between the energy parameters considered above is noted:

$$E_a^\beta \approx (0.3 \pm 0.05) E_c + B. \tag{2.4}$$

In this case, E_a^β and E_c are assigned to 1 mole of segments, and the term $B \approx 10 \pm 5$ kJ/mol corresponds to the barrier of internal rotation in flexible-chain polymers. Consequently, in the polymers, the activation barrier of the β-transition is also determined by the energy of intermolecular interactions, as for the low-molecular-weight bodies (expression (2.2)). The difference consists in taking into account the term B, which expresses the role of conformational rearrangements in the elementary act of β-relaxation.

Thus, the data considered allow us to formulate a unified concept of the relaxation β-process mechanism for flexible-chain polymers. This mechanism includes the rotational and translational movements of segments with the overcoming of intermolecular activation barriers with the participation of a conformational, presumably, *trans-gauche* transition.

A similar act is analogous to an elementary act of viscous flow due to segmental mobility and is realized in a solid polymer in local, liquid-like structural regions with a reduced density and an excess of free volume [20].

From this point of view, the temperature T_β is given the meaning of the lower boundary temperature, from which the processes due to local segmental mobility, namely, physical ageing, thermostimulated relaxation of latent energy and dimensions of deformed samples, plasticity, etc., appear in the polymer. These issues are discussed in more detail in the relevant chapters and sections of this monograph.

2.1.2. Relaxation α-transition and vitrification

A relaxation α-transition of the non-Arrhenius type is the main relaxation transition for the amorphous polymers (Fig. 2.1) or, in general, for the disordered bodies. A sharp change in the physical, thermophysical, and physico-mechanical properties of the material is observed in this temperature range (Fig. 2.2).

It is obvious that the α-relaxation is associated with the 'thawing' of the mobility of a certain type of kinetic units. For low-molecular-weight glassy liquids, this unit is a molecule whose mobility is responsible for vitrification (in this context, the α-transition)[1], and for the viscous flow. As a result, when the glass transition temperature T_g, which coincides with the flow temperature T_f, is reached, the low-molecular-weight glass transforms into a viscous-flow state.

---

1) The correspondence between the concepts 'relaxation α-transition' and 'vitrification' is discussed below.

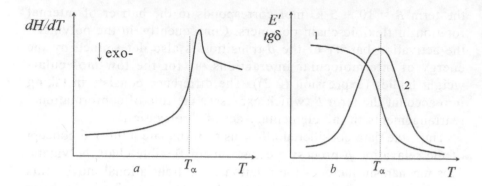

Fig. 2.2. A typical DSC curve (*a*) and typical temperature dependences of the elastic modulus E' (1) and the loss tangent tg δ (2) (*b*) of an amorphous polymer.

For linear flexible-chain polymers, the kinetic unit determining the relaxation α-transition is a segment whose thermal activation at T_α causes the amorphous polymer to transition to a high elastic state. The development of the viscous flow and the transition to the viscous flow state requires the 'thawing' of the mobility of macromolecular coils, which occurs at a flow temperature which for polymers is much higher than T_g and T_α (see Section 1.3).

In the middle of the last century, the question of the cooperative nature of the α-transition, associated with correlated, coordinated displacements of several segments, was discussed [31–35]. It should be borne in mind that the correlation of segmental mobility can have both an intrachain and an interchain character.

The results of the cycle of studies, generalized in the monograph [20], indicate that the size of the kinetic unit responsible for the manifestation of α-relaxation is comparable to the size of the statistical segment determining the relaxation β-transition. This allowed us to conclude that the 'large-scale' α-transition in comparison with the β-transition ($E_a^\alpha \gg E_a^\beta$) is due not to the increase in the size of the kinetic unit, as would be expected in the case of the co-operative mobility of several segments, but to the correlated displacement of the contacting segments of neighbouring chains.

Thus, on the basis of the foregoing, we can conclude that the relaxation α-transition is due to the cooperative modes of segmental mobility, which in the β-transition are realized as quasi-independent motions localized in structural regions with a reduced packing density.

The inter-relation between the α- and β-transitions for the disordered bodies of both polymeric and non-polymeric nature can be clearly seen in the correlation of their activation energies

$$\left(E_a^\alpha / E_a^\beta \approx 4 \pm 1 \right) \tag{2.5a}$$

and the Boyer ratio [36, 37]

$$\left(T_\beta / T_\alpha \approx 0.75 \pm 0.1 \right). \tag{2.5b}$$

An additional indication of their unity can be the above-mentioned fusion of these two transitions into a common $\alpha\beta$-transition at high effect frequencies, as well as mutual $\alpha-\beta$-transformations recorded for some systems by differential scanning calorimetry [20]. In both cases, the observed effects are determined by the manifestation or degeneration of the cooperative segmental movement.

The relaxation α-transition is usually identified with the vitrification of an amorphous body. In fact, the glass transition is a more complex phenomenon. As shown above with the example of α- and β-relaxation, the relaxation transitions are associated only with thermal activation ('thawing' – 'freezing') of some form of molecular mobility without appeal to a change in the structural state of the sample. Vitrification also includes both the molecular–kinetic and structural aspects. To explain this thesis, let us consider the temperature dependence of the specific volume[2] of the amorphous polymer V_{sp} (Fig. 2.3).

This dependence is characterized by a pronounced kink at glass transition temperature T_g, i.e. at the transition temperature of the amorphous polymer from the vitreous to the high elastic state. At the same time, the coefficient of volumetric thermal expansion increases sharply (by a factor of 3–4). To explain the observed behaviour, we will use the concept of free volume.

In the simplest version of this approach, the free volume V_{fr} is introduced as the difference between the macroscopic volume V and the occupied volume V_{oc}, i.e. the total volume of molecules constituting the given body:

$$V_{fr} = V - V_{oc}.$$

---

[2]Specific volume is the volume of a unit of mass of a substance. The inverse of the density.

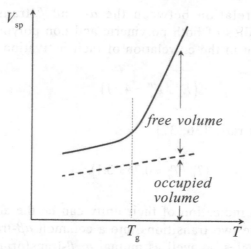

Fig. 2.3. A typical temperature dependence of the specific volume V_{sp} for an amorphous polymer.

The occupied volume increases linearly with increasing temperature (dashed line in Fig. 2.3). At $T < T_g$ for the amorphous flexible-chain polymers, the free volume fraction is constant and amounts to 0.025–0.03 of the total volume of the sample. At $T > T_g$, a noticeable increase in the fraction of free volume is observed in the polymer with increasing temperature.

This interpretation allows us to conclude that the vitrification of an amorphous polymer as it cools occurs when the fraction of the free volume reaches a critical value of 0.025–0.03. At T_g, this structural state 'freezes' and remains unchanged at temperatures below the glass transition temperature, i.e. in the temperature range of the existence of a polymer glass. Conversely, when the glassy polymer is heated, its 'devitrification', i.e. transition to a highly rubbery state occurs when the fraction of the free volume exceeds the above critical value.

Thus, vitrification is distinguished by a dualistic nature.

When cooled, this transition is realized as a result of the simultaneous flow of two processes:

1. 'freezing' of the cooperative movement of segments (molecular-kinetic aspect, described in terms of α-relaxation);

2. 'freezing' of a strictly defined (0.025–0.03) fraction of the free volume (structural aspect).

'Devitrification' as a result of polymer heating requires simultaneous

1. thermally activated 'thawing' of cooperative segmental mobility;
2. appearance in the polymer of an adequate fraction of the free volume $V_{fr} > 0.025\text{--}0.03$.

The relationship between these two events is obvious – the thermally activated displacement of the kinetic unit responsible for the vitrification is possible only if there is a neighbouring free volume 'hole' (Fig. 1.17a).

In fact, the concept of 'vitrification' has an extremely broad meaning, often specific in describing certain physical, physico-chemical or chemical processes.

For example, under the conditions of dynamic mechanical analysis at a constant temperature, the increase in the frequency of mechanical action is accompanied by the so-called 'mechanical' vitrification of an amorphous body (Fig. 2.4). A detailed analysis of the observed behaviour is given in Appendix 1. In the context of this section, we only briefly mention the following considerations.

The constancy of the test temperature means the constancy of the relaxation time τ of the kinetic unit, the mobility of which determines the manifestation of this transition. The variable parameter in this case is the time of mechanical action t inversely proportional to the frequency ω: $t \sim 1/\omega$.

At low frequencies, the value of t is large and considerably exceeds τ: $\tau \ll t$. Under these conditions, the kinetic units manage

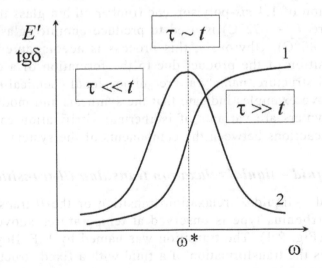

Fig. 2.4. Typical frequency dependences of the elastic modulus E' (1) and loss tangent tg δ (2) of the amorphous body.

to regroup, and the amorphous body demonstrates a liquid-like (or, in the case of polymers, a high elastic) response to the applied mechanical action.

As the frequency increases, the value of t decreases, and at high frequencies the ratio $\tau \gg t$ is realized, which determines the solid-state response characteristic of the vitreous body. As a result, the elastic modulus E' increases by 2–3 orders of magnitude (curve *1*, Fig. 2.4).

The boundary condition for such an isothermal, mechanically stimulated vitrification is the ratio $\tau \sim t$, which is characteristic of the boundary critical frequency ω^*.

Vitrification is also an essential feature of polymer synthesis from a liquid monomer. For example, in the simplest version of bulk radical polymerization at a fixed temperature, the growth of macromolecules causes an increase in the viscosity of the liquid polymerization system and a regular increase in the glass transition temperature. At a certain conversion, the glass transition temperature of the polymerization system reaches the polymerization temperature, and vitrification occurs. The glass transition of the system means a kinetic inhibition of the polymerization due to the fact that in the glass the translational mobility of the monomer molecules is 'frozen'. To ensure further polymerization, a rise in temperature is required.

A similar situation is observed when the linear polymers are crosslinked by chemical and radiation methods. For example, the vulcanization of 1,4-*cis*-polyisoprene (rubber with a glass transition temperature T_g = –72°C) is used to produce ebonite (glassy resin with T_g = 80°C). Obviously, this process is accompanied by the glass transition of the product due to the formation of a complex topological structure, namely a three-dimensional chemical network.

The above examples indicate that the synthesis and modification of the polymers are variants of isothermal vitrification caused by chemical reactions between the components of the system.

2.1.3. 'Liquid – liquid' relaxation transition (ll-transition)

The 'liquid 1–liquid 2' relaxation transition or the *ll*-transition of the non-Arrhenius type is observed at temperatures above the α-transition (Fig. 2.1). The transition was named by R.F. Boyer [17] and denotes the transformation of a fluid with a fixed structure into an unstructured 'true' liquid.

The experimental material [20,38–42] was accumulated already at the end of the last century, clearly determining for a wide range of flexible-chain polymers the relation between the temperatures of the α-transition T_α and the *ll*-transition T_{ll}:

$$T_{ll} / T_\alpha \approx 1.2 \pm 0.05. \qquad (2.6)$$

It was found that at T_{ll} the polymers show

- a sharp increase in the interchain distance, i.e. decrease in intermolecular ordering [3];
- a noticeable decrease in the intermolecular interaction energy [44[;
- a twofold decrease in the activation energy of molecular mobility [38– 41] in comparison with the α-transition:

$$\left(E_a^\alpha / E_a^{ll} \approx 2 \pm 0.05 \right); \qquad (2.7)$$

• abrupt change in ultrasound velocity [45].

In the monograph [20], the temperature of the *ll*-transition in amorphous, non-crystallizing polymers is correlated with the temperature of the maximum crystallization rate of the crystallizing polymers T_{cr}^{max} (Fig. 2.5).

When cooling from the melt for crystallizing substances of both polymeric and non-polymeric nature, relation $T_{cr}^{max} \approx (0.8–0.85) T_m$ [4,6,46] is satisfied.

Taking into account the known 'two-thirds' rule $T_g/T_m \approx 2/3$, we obtain the following relation:

$$T_{cr}^{max} \approx (1.25 \pm 0.05) T_g \approx T_{ll},$$

which indicates an analogy in the behaviour of crystallizable and non-crystallizing polymers upon cooling of their melts.

The set of the given data can be interpreted in terms of the association of kinetic units (molecules, segments) in the liquid phase [20, 47–49], which causes the formation of fluctuation, dissipative structures. The resulting associates are characterized by the distribution by the number of kinetic units included in the associate, packing density, lifetime, etc. The specificity of polymers due to their chain nature is that the segment associates (clusters, supersegmental λ-structures) play the role of nodes of the physical, fluctuation network (see Appendix 2).

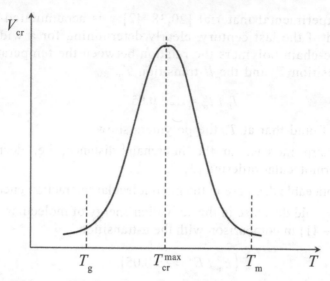

Fig. 2.5. Temperature dependence of polymer crystallization rate.

When the polymer liquid is cooled to T_{ll}, the conditions for the appearance of kinetically stable[3] clusters or λ-structures are realized. In the case of crystallizable polymers, the associates formed serve as embryos of crystallization, ensuring the maximum process speed at a given temperature (Fig. 2.5). The growth of crystallites is accompanied by the formation of a crystalline phase characterized by a long-range three-dimensional order. For non-crystallizable polymers, the local ordering of the segments stops at the stage of clustering with the formation of λ-structures, i.e. short-range order regions. A further decrease in temperature leads to an increase in the number of clusters, as well as an increase in their defectiveness and a decrease in density. These effects should be associated with the fact that the clusters that have already arisen (nodes of the physical network) limit the mobility of intercluster, tie-chains and prevent the segments from being placed in the constantly forming supersegmental structures. At T_g the mobility of the segments ceases and the fluctuation, dissipative structure, which is a cooperative system including short-range regions (clusters or λ-structures) and more loose intercluster interlayers, 'freezes' in the polymer.

Heating of the amorphous polymer from T_g to T_{ll} leads to the decay of more and more perfect and long-lived clusters, and at T_{ll} a

---

[3]Kinetically stable are structural elements whose lifetime is much longer than the time of the experiment.

transition to a 'true' liquid with a completely destroyed dissipative structure is observed.

For real systems, the relaxation λ-processes associated with the destruction of the nodes of the physical network (λ-structures) and responsible for the *ll*-transition are more complex. The spectra of internal friction of a number of amorphous polymers [21] clearly resolve multiplet λ-transitions in temperature ranges

- $(1.3–1.5)\ T_\alpha$ – poly(methyl methacrylate);
- $(1.15–1.4)\ T_\alpha$ – polystyrene;
- $(1.3–1.6)\ T_\alpha$ – polydimethylsiloxane;
- $(1.4–1.6)\ T_\alpha$ – polybutadiene;
- $(1.1–1.4)\ T_\alpha$ – butadiene-acrylonitrile elastomer SKN-40;
- $(1.4–1.8)\ T_\alpha$ – butadiene-methylstyrene elastomer SKMS-10.

The intervals of manifestation of the relaxation λ-processes correlate quite well with the relation (2.6), and the multiplicity of λ-transitions is obviously related to the distribution of clusters or λ-structures noted above in size, defectiveness, packing density, etc.

2.1.4. On the relationship of β-, α-, and ll-transitions

First of all, we give the following correlations and experimental results.

- Comparison of expressions (2.5*b*) and (2.6) indicate a quantitative relationship between the temperatures of all three transitions:

$$T_\beta \approx 0.75 T_\alpha \approx 0.62 T_{ll}.$$

- The activation energies of the β-transition (E_a^β), the viscous flow (E_a^η), the α-transition (E_a^α) and the *ll*-transition (E_a^{ll}) are found in the following relationship (see expressions (2.2)–(2.4), (2.5 *a*) and (2.7)):

$$E_a^\beta \approx E_a^\eta \approx E_a^\alpha / 4 \approx E_a^{ll} / 2.$$

Schematically, this situation is depicted in Fig. 2.6.

- For polymerhomological series[4], the temperature dependences of the considered transitions on the degree of polymerization P_n tend to reach a constant value when a certain critical value

---

[4]Polymerohomological series is a series of chemical compounds of chain structure and the same chemical nature with a gradually increasing degree of polymerization or molecular weight.

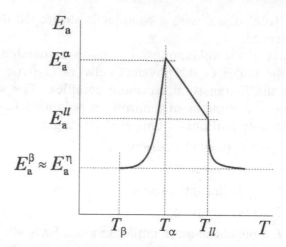

Fig. 2.6. The temperature dependence of the effective activation energy of molecular mobility in an amorphous polymer [20].

P_n^* is reached (Fig. 2.7) [20, 40, 41, 50, 51]. This means that the manifestation of all three transitions is associated with the thermal activation of the same kinetic unit, comparable in scale to the segment.

- For all three transitions, the decrease in the intermolecular interaction energy [44] and the increase in the interchain distance are recorded, i.e. decrease in intermolecular ordering [43].

The set of obtained data allows us to conclude that the elementary kinetic unit determining the relaxation behaviour of a flexible-chain amorphous polymer in the temperature range from T_β to $T > T_{ll}$ is a segment of a macromolecule. The differences in molecular dynamics in the vitreous (at $T = T_\beta - T_\alpha$), high elastic (at $T = T_\alpha - T_{ll}$) and viscous flow (at $T > T_{ll}$) states are due only to the ratio of the kinetically independent and correlated acts of segmental mobility.

At $T \sim T_\beta$ and at $T > T_{ll}$, independent β-processes of the Arrhenius type predominate and their activation energy is comparable to that of a viscous flow (Fig. 2.6). When the polymer is cooled from T_{ll} to T_α, the contribution of cooperative α-processes increases, reaching a maximum at T_α. Further lowering of temperature gradually 'freezes' the cooperative segmental mobility, and at T_β the independent movement of the segments is localized in liquid-like structural regions with a reduced packing density and an excess of free volume.

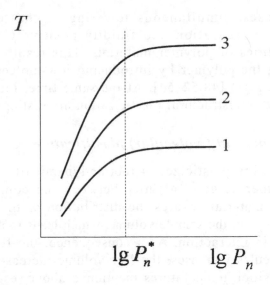

Fig. 2.7. Typical dependences T_β (1), T_α (2) and T_{ll} (3) for polymerhomological series.

2.2. Plasticized polymers

The response of the amorphous polymer to the applied mechanical loading is determined by the ratio of the operating temperature T_{op}, on the one hand, and the brittleness temperature T_{br}, the glass transition temperature T_g, and the flow temperature T_f, on the other hand.

At $T_{op} < T_{br}$, the polymer behaves like a brittle glassy body. In this temperature range, the set of properties (first of all, plasticity) for the polymers is lost, favourably distinguishing them from low-molecular-weight glassy materials. In this regard, below the temperature of brittleness, the polymers are of no interest from an operational point of view.

When $T_{br} < T_{op} < T_g$, the deformation of amorphous polymers proceeds according to the yielding mechanism. This determines the most valuable feature of polymer glasses, and, in particular, their high plasticity.

In the temperature range $T_g < T_{op} < T_f$, the amorphous polymer is in a high elastic state and has the capacity for large reversible deformations, which is an inherent consumer characteristic of elastomers.

At the flow temperature T_f, the amorphous polymer transforms into a viscous flow state which is mainly used for polymer processing.

In many cases, simultaneous lowering of the temperature of brittleness, vitrification and fluidity positively affects the consumer qualities of polymer materials. This result is achieved by plasticizing the polymer by introducing low-molecular-weight liquids (plasticizers) [48,52–54]. At the same time, intrastructural (molecular) and interstructural plasticization are distinguished.

2.2.1. Intrastructural (molecular) plastification

In this variant, the plasticizer is thermodynamically compatible with the polymer. A good affinity between the components of the plasticized material ensures the distribution of the plasticizer molecules throughout the sample volume, which leads to a weakening of the interchain interaction. As a consequence, the flexibility of the macromolecules increases, the free volume increases, and as a result, the transition temperatures mentioned above, especially the glass transition temperature [55–57], decrease.

Intrastructural plastification is characteristic for two types of systems: 'polar polymer – polar plasticizer' and 'nonpolar polymer – nonpolar plasticizer'.

Strong intermolecular interactions, characteristic of polar polymers, determine appreciable limitations of chain mobility and an effective increase in their rigidity. This naturally causes relatively high glass transition temperatures of these materials. The molecules of a polar plasticizer compatible with a polymer penetrate into the interchain space, solvate the polar groups of macromolecules, thereby reducing the energy of intermolecular interactions and increasing the flexibility and mobility of macrochains. As a result, a decrease in T_g of the material is observed. In other words, the plasticizing effect of a polar plasticizer on a polar polymer is of an energetic nature.

Quantitatively, the plasticization efficiency is estimated from the value of the parameter ΔT_g, which is the difference in the glass transition temperature of the initial and plasticized polymer.

For the systems 'polar polymer–polar plasticizer', the Zhurkov rule or the rule of mole fractions has the form [58]:

$$\Delta T_g = kn,$$

where n is the mole fraction of the plasticizer, k is the coefficient specific for this 'polymer–plasticizer' pair.

When a nonpolar polymer is plasticized with a nonpolar plasticizer, the main role is played not by energy but by entropy factors – the larger the volume filled with plasticizer molecules, the greater the chain mobility and, consequently, the lower the glass transition temperature of the plasticized sample.

For the systems 'nonpolar polymer–nonpolar plasticizer' Kargin and Malinsky proposed the rule of volume fractions [59]:

$$\Delta T_g = k'\varphi,$$

where φ is the volume fraction of the plasticizer, k' is the coefficient characteristic for the given 'polymer–plasticizer' pair.

The concentration range in which the rules for molar and volume fractions are satisfied is determined by the compatibility of the polymer and the plasticizer. For 'polymer–plasticizer' systems characterized by unrestricted compatibility of components, the glass transition temperature of the plasticized material is practically linearly reduced to the glass transition temperature of the plasticizer. In case of limited compatibility, as the plasticizer concentration increases, the glass transition temperature of the polymer reaches its constant value, and does not change with a further increase in the plasticizer content.

Note that from the practical point of view, plasticizers with low volatility and low diffusion coefficients in the polymer matrix are chosen to increase the life of plasticized materials. These two factors prevent 'sweating' of the plasticizer from the product, increasing the stability of its performance properties. Typical plasticizers for industrial polymers are dibutyl phthalate, dioctyl phthalate, tricresyl phosphate and the like, as well as hydrogenated monomers. For hydrophilic polymers, water is an effective plasticizer.

In the general case, for amorphous linear polymers, plasticization leads to a decrease in both the glass transition temperature and the flow temperature, practically without affecting the width of the high-elasticity temperature range $T_f - T_g$.

For crosslinked rubbers, the region of high elastic deformations lies between the glass transition temperatures and the thermal destruction of the material. In this case, the introduction of plasticizers reduces only the glass transition temperature, thereby widening the temperature interval for the operation of the rubber.

The effect of molecular plasticization on the physico-mechanical behaviour of polymers is much more complicated than the above mentioned reduction in the transition temperatures, primarily the glass transition temperature. In fact, this method of physical and chemical modification of polymers radically changes the relaxation spectrum of the material. The quantitative parameters of such changes vary for specific pairs of 'polymer–plasticizer' and are reflected in the original articles. Here we shall distinguish the main features of these phenomena [20].

The method of differential scanning calorimetry showed, firstly, that in the case of intrastructural plasticization, the jump in the specific heat at the glass transition temperature of the plasticizer is not observed, which means that there is no microphase separation in the plasticized systems.

Secondly, the effect of the plasticizer on the relaxation behaviour of the polymer is reduced to the following trends:

1. displacement of T_α in the region of lower temperatures, broadening of the α-transition interval, and a decrease in the heat capacity jump for a given transition;

2. the fusion of the β- and α-transitions – $T_\alpha \rightarrow T_\beta$, $E_a^\alpha \rightarrow E_a^\beta$, where E_a are the activation energies of the corresponding transitions;

3. appearance of intermediate relaxation transitions in the temperature interval between T_β and T_α.

The observed effects associated with changes in molecular dynamics have a significant effect on the mechanical properties of polymers. Plasticization leads to a decrease in the modulus of elasticity and the yielding stress of the material, as well as an increase in its elasticity and elongation at break. The same effect on the mechanical behaviour of the initial, unplasticized polymer is caused by temperature increase (see Section 4.1).

An analogy of the effect of plasticization and temperature on the mechanical parameters of a polymer can be easily shown as follows [15, 60].

Thus, at a given deformation temperature T_{def}, an increase in the concentration of the plasticizer is accompanied by a near-linear decrease in any mechanical characteristic of the polymer, for example, the yielding stress σ_y by decreasing the glass transition temperature and, consequently, softening the material. For a given unplasticized polymer with a certain glass transition temperature, a linear decrease in σ_y is achieved by increasing T_{def}.

Let us introduce the relative deformation temperature $\Delta T = T_g - T_{def}$. Obviously, this temperature can be lowered in two

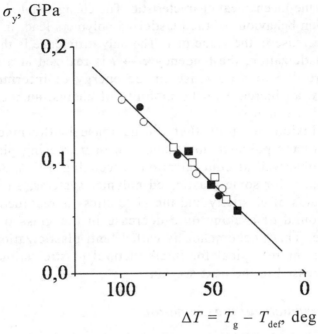

Fig. 2.8. Dependence of the yield stress σ_y on the relative deformation temperature ΔT for poly(methyl methacrylate) (light circles), poly (methyl methacrylate) plasticized with dibutyl phthalate, (dark circles), poly(vinyl chloride) (light squares), and poly(vinyl chloride), plasticized with dioctyl phthalate (dark squares).

ways: either by decreasing T_g for a fixed T_{def} (plasticized samples), or by increasing T_{def} for the initial polymer with a fixed T_g. Both cases satisfy the universal dependence in the co-ordinates $\sigma_{f.e.} = f(\Delta T)$ (Fig. 2.8). Similar universal correlations have also been obtained for other mechanical characteristics – the elastic modulus yield strain ε_y.

This result clearly indicates a temperature–concentration analogy, meaning that the same mechanical response of the polymer can be achieved either by plasticization or by a change in temperature.

The temperature–concentration analogy for plasticized systems is a consequence of the temperature–time superposition (see Section 1.4). Recall that the deformation behaviour of the polymer, and, hence, the mechanical parameters of the material are determined by the ratio of the time of action t, which is determined by the deformation rate, and the relaxation time of the kinetic unit responsible for deformation, τ. The results shown in Fig. 2.8, were obtained at a fixed rate of deformation, which means constancy of t. As the deformation temperature increases, the relaxation time of the kinetic unit, in our case, of the segment decreases ($\tau \rightarrow t$), which causes a decrease in the mechanical characteristic. The changes noted above in the relaxation behaviour of the plasticized polymers lead, in general, also to a decrease in the value of τ. The only difference is that in the case of plasticization, the tendency $\tau \to t$ is realized at a constant temperature due to a decrease in the energy of intermolecular interactions, an increase in the mobility of chains, an increase in the free volume, and so on.

In conclusion, we note that the decrease in the mechanical properties of the polymer noted above with increasing plasticizer content is observed at concentrations exceeding 3–5%. At lower concentrations, for some plasticized polymer systems, an increase in the modulus of elasticity and the yield stress is recorded against the background of a monotonic decrease in the glass transition temperature. This phenomenon is called 'anti-plasticization'. This behaviour is more typical for interstructural plasticization, which will be discussed in the next section.

2.2.2. Interstructural plasticization

Interstructural plasticization is based on introduction into the polymer of a plasticizer having a low thermodynamic affinity for the matrix [54[. In this case, the uniform distribution of the plasticizer molecules throughout the sample volume does not occur, and the plasticizer concentrates in places with a reduced packing density at the boundaries of the supersegmental structural formations (see Appendix 2). In other words, the interstructural plasticizer acts not at the level of interchain contacts, but acts as a 'lubricant' when moving these nanometric structures. Let us consider the basic laws of interstructural plasticization on a concrete system, which reflects a rather general character of this phenomenon.

The physico-mechanical and relaxation behaviour of poly(methyl methacrylate) (PMMA), plasticized with diethylsiloxane liquid (DES), was studied in Refs. [12,61,62]. This ingredient is readily soluble in the monomer (methyl methacrylate), but insoluble in PMMA. The diethylsiloxane liquid was introduced into the monomer, and as the polymerization progressed, the degradation of the affinity of the DES to the polymer formed caused it to be 'displaced' in a region with an excess of the free volume.

Figure 2.9 shows the dynamometric curves of polymer samples with different contents of DES. It is clearly seen that the introduction of 0.1% DES leads to an increase in the modulus of elasticity and

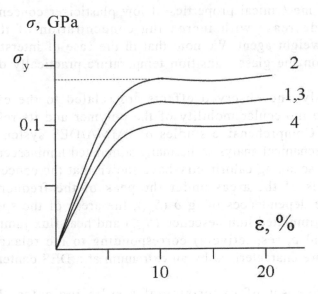

Fig. 2.9. The stress–strain diagram of the initial PMMA (1) and PMMA containing 0.1 (2), 0.5 (3), and 1.3% of DES (4).

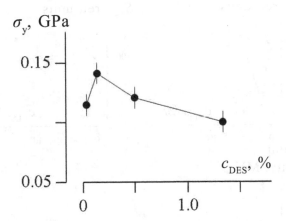

Fig. 2.10. Dependence of the yield stress σ_y on the concentration of DES in the PMMA.

the yield stress of the polymer, while a further increase in the concentration of DES in the polymer is accompanied by a decrease in these mechanical parameters. For the yield stress σ_y this trend is shown in Fig. 2.10.

Such extremal dependences are described in terms of 'anti-plasticization' of the polymer, a phenomenon associated with an increase in mechanical properties at low plasticizer concentrations and their decrease with increasing concentration of the low-molecular-weight agent. We note that in the case of interstructural plasticization the glass transition temperature practically does not change.

Obviously, the observed effects are related to the effect of DES on the molecular mobility of the polymer and its relaxation behaviour. Comprehensive studies of PMMA/DES systems using dynamic mechanical analysis, thermally stimulated luminescence, and differential scanning calorimetry have shown that the concentration dependences of the areas under the peak of the frequency and temperature dependences of tg δ ($S_{tg\delta}$), the areas of the spectra of thermally stimulated luminescence (S_{TSL}) and heat flux jumps (Figs 2.11a, b and c, respectively) corresponding to the relaxation α-transition, are characterized by an extremum at a DES content equal to 0.1%.

The above set of experimental results indicates that the introduction of a 0.1% plasticizer into the polymer leads to a

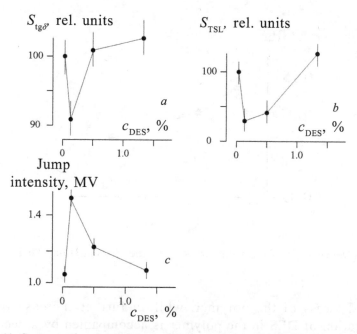

Fig. 2.11. Concentration dependences of the parameters of the relaxation α-transition in the PMMA/DES system: the peak area of the frequency and temperature dependences of the loss tangent tg δ (a), the area of the spectrum of the thermally stimulated luminescence S_{TSL} (b), and the intensity of the heat flux jump (differential scanning calorimetry) (c).

decrease in the mobility of the segments, and a further increase in the concentration of DES – to an increase. Of course, the above conclusion is correctly applied to molecular dynamics only in the temperature interval of the α-transition. However, similar extreme concentration dependences of the mechanical characteristics of the material (Fig. 2.10) suggest that this trend persists even at temperatures below T_g.

For a deeper understanding of the nature of interstructural plasticization, let us turn to the results obtained by positron annihilation.

The positron annihilation method is widely used to study the evolution of free volume in polymers under various physical and physico-chemical influences [12,61,63–69]. The fixed experimental parameters (long-lived components of the spectra – the lifetime of positronium τ_L and the intensity of positronium annihilation I_L), at least qualitatively, characterize the effective size of the fluctuation pores and their concentration, respectively. The product of the annihilation characteristics $\tau_L \times I_L$ has the meaning of a fluctuation free volume.

Figure 2.12 shows the concentration dependences of the long-lived components of the annihilation spectra for PMMA, plasticized with DES. The data presented allow us to conclude that the introduction of 0.5% and 1.3% of DES in the polymer is accompanied by a noticeable increase in the fluctuation free volume of the material. With a DES of 0.1%, there is a certain tendency to reduce the free volume or, at least, to ensure its constancy.

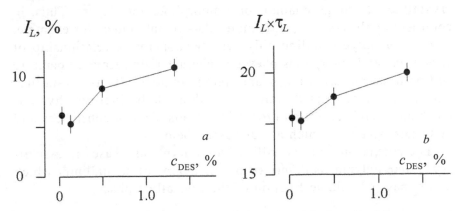

Fig. 2.12. Concentration dependences of the intensity of the long-lived I_L component of the positron annihilation spectrum (*a*) and the product of the intensity of a long-lived component for its lifetime τ_L (*b*) for PMMA/DES.

The results obtained indicate that the introduction of a diethylsiloxane liquid with a concentration of 0.1% into the polymer does not lead to a noticeable change in the free volume recorded by the positron annihilation method. This may be due to the fact that during the synthesis process, the liquid incompatible with the polymer is concentrated in cavities of free volume, continuously formed as the polymerization proceeds at the boundaries of the superegmental structures (clusters) (see Appendix 2). These intercluster structural levels are sufficiently large and are located at distances much greater than the positron and positronium range. In connection with this, the detection of such regions lies beyond the limits of the possibilities of the method [70–72]. The introduced liquid limits the molecular mobility in the vicinity of these structural formations, which is expressed macroscopically in the change in the intensity of the relaxation α-transition (Fig. 2.11) and the growth of the mechanical parameters of the obtained samples (Fig. 2.10).

With an increase in the content of DES to 0.5 and 1.3%, this ingredient loosens the structural regions with a reduced packing density, increasing the intensity of segmental mobility in them and thus reducing the mechanical characteristics of the material.

2.3. Semi-crystalline polymers

A specific feature of semi-crystalline polymers is the coexistence of crystalline and amorphous phases. During crystallization, the same macromolecule participates in the formation of several crystalline regions (crystallites). The segments of the tie-chains connecting the crystallites, (through chains) form amorphous interlayers. Thus, in contrast to polycrystalline low-molecular-weight bodies, for example, metals for polycrystalline polymers, a cooperative relationship of crystalline and amorphous phases is characteristic. Any temperature and/or force perturbation of amorphous regions affects crystallites and vice versa. This dictates a number of distinctive physico-mechanical and relaxation properties of this class of substances and materials, some of which are discussed below.

The coexistence of crystalline and amorphous phases in a semi-crystalline polymer is described by the degree of crystallinity which is the mass or volume fraction of the crystalline phase:

$$\chi = \frac{m_{cr}}{m} = \frac{\rho_{cr} V_{cr}}{\rho V},$$

where m and m_{cr} are the mass of the sample and its crystalline phase; ρ and ρ_{cr} are the density of the sample and its crystalline phase; V and V_{cr} are the volume of the sample and its crystalline phase.

Experimentally, the degree of crystallinity is determined using differential scanning calorimetry, X-ray diffraction analysis, infrared spectroscopy, etc.

The structure of semi-crystalline polymers (see Appendix 3) is more complex and regular than the structure of amorphous samples and can be quantitatively characterized using a number of experimental techniques, primarily X-ray diffraction analysis. It would seem that the limitation of various forms of molecular mobility as a result of crystallization should be accompanied by a depletion of the relaxation spectrum of the polymer. In practice, however, the picture in the overwhelming majority of cases is directly opposite – the number of relaxation transitions in semi-crystalline polymers is much greater, for example, from dynamic mechanical analysis and differential scanning calorimetry, than for amorphous polymer glasses.

These observations relate [20, 21] to a wide range of relaxing kinetic units, including structural elements

- inside crystallites;
- in intercrystalline amorphous regions;
- in the interphase layer on the surface of the crystallites.

In more detail, we consider this situation using the example of a lamellar structure of polyethylene (PE).

In the monograph [73], a fine structure of this polymer is proposed (Fig. 2.13). The following structural kinetic units are distinguished:

1. between the crystalline cores of the lamellas;
2. regular folds with suppressed mobility;
3. irregular loops;
4. folded tie-chains;
5. free ends of macromolecules coming out of lamellas;
6. slightly curved tie-chains;
7. folds the mobility of which is significantly limited by crystallites;
8. fully straightened tie-chains, the ends of which are fixed by neighbouring lamellas.

A semi-quantitative estimate of the mobility of the indicated kinetic units can be their degree of convolution $c = l/h$, where h is the distance between the ends of a given element, and *l* is its contour length. Obviously, the higher *c* of a given structural unit, the lower its temperature at which its mobility is 'thawed'.

We note that for specific samples the type of experimentally obtained relaxation spectra depends on the chemical nature of the polymer, the conditions for its crystallization, the temperature-temporal prehistory, etc., which determines the degree of crystallinity, morphology, and the fine structure of the crystalline and amorphous regions of the final material. In other words, the relaxation spectrum is specific not only for a given polymer, but also for a specific sample of the polymer. A detailed description of the observed diversity of experimental data goes beyond the scope of this paper. Here, for the example of PE, we consider the general picture of the relaxation behaviour of a semi-crystalline polymer, which is typical for other representatives of this class of materials.

Investigations of the relaxation phenomena in PE using various experimental techniques (differential scanning calorimetry, dynamic mechanical analysis, nuclear magnetic resonance, etc.) make it possible to distinguish in the polymer the following types of relaxation processes:

- Relaxation I in the temperature range 140 – 170 K;
- Relaxation II (240–270 K);
- Relaxation III (300–370 K);
- Relaxation IV (385–400 K), which precedes melting and is observed at 10–15° below the melting point.

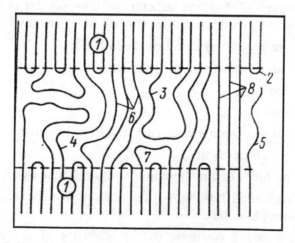

Fig. 2.13. Schematic representation of the fine structure of lamellar PE [73]. Explanations in the text.

Let us correlate these types of relaxation with certain elements of the structure of a semi-crystalline polymer (Fig. 2.13).

First of all, we note that removal of the amorphous component of the semi-crystalline polymer, for example, by etching the samples, leads to the complete disappearance of peaks corresponding to relaxation of types I–III [74]. This makes it possible to attribute with certainty the manifestation of these relaxation phenomena of mobility of structural units in intercrystalline regions, i.e. elements 3–7 (Fig. 2.13).

Based on nuclear magnetic resonance data [75[, the following groups of relaxing kinetic units are distinguished in the amorphous regions of PE:

- **Group 1**, which consists of relatively straightened chains, for which the degree of curvature c does not exceed 1.4;

- **Group 2**, which includes a set of structural units with a spread c ranging from 1.5 to 4.0.

We note that for interlamelar regions of unoriented PE the fraction of the kinetic units of group 1 naturally increases with the degree of crystallinity of the polymer.

From the point of view of molecular dynamics, the elements of group 2, primarily irregular loops (3), convoluted tie-chains (4), and sufficiently long free ends of macromolecules (5), (Fig. 2.13) have a wide set of conformations, comparable practically with the conformational set of the statistical coils [75,76].

The simulation of the behaviour of these kinetic units in the interlamelar space by the Monte Carlo method [77] shows that their contour lengths can exceed the thickness of the amorphous interlayer by a factor of 1.5. Similar values of these quantities are also noted in [20,74]. In other words, the linear dimension of such structural formations is comparable to the length of the Gaussian subchain and includes several tens of monomer units.

In view of this, the mobility of the elements of group 2 depends little on the presence of crystallites and appears at rather low temperatures in the relaxation region I and II. The nature and mechanism of these transitions are assumed to be [20] analogous to those for the β- and α-relaxation of amorphous polymers (see Section 2.1). Recall that as an elementary kinetic unit responsible for these transitions, we consider the segment of the macromolecule, where the β-relaxation (relaxation I) is realized due to quasi-independent, localized displacements of the segments, and for α-relaxation (relaxation II), these quasi-independent modes acquire a cooperative, correlated character. This conclusion is supported by quantitative estimates of the activation parameters of these transitions, as well as the fact that relaxation I refers to the Arrhenius processes, and relaxation II to non-Arrhenius processes [20]. From these positions, the relaxation II is correlated with the vitrification of the amorphous component of the semi-crystalline polymer.

Relaxation II is clearly observed only in samples with a rather low degree of crystallinity (25–30%). The increase in the fraction of the crystalline phase to 50% and higher completely suppresses this transition, practically without affecting the relaxation I.

Relaxation III is associated with the 'defrosting' of segments in the elements of group 1 – slightly curved transition chains (6) and folds (7), the dynamics of which are markedly limited by the crystallites (Fig. 2.13) [20]. The temperatures of these transitions and their activation parameters increase with increasing degree of crystallinity and thickness of the lamellae. Obviously, an increase in the fraction of the crystalline phase is accompanied by a decrease in the degree of convolution of these subchains and, as a result, by a restriction of their mobility, which determines the observed effects.

Relaxation IV is attributed [75] to the appearance of segmental mobility in fully straightened tie-chains (8) (Fig. 2.13). Ambiguity of the interpretation of this transition is due to the fact that, on the one hand, the 'bundle' of maximally rectified, parallel-laid sections of macromolecules is the nucleation of crystallization. In connection with this, the 'defrosting' of the mobility of these elements leads to a spontaneous crystallization of such a structural preform, which, naturally, distorts the relaxation spectrum. On the other hand, the thermal activation of segmental mobility in these tie-chains (8) is largely determined by the phase state of the lamellas, in which the ends of these structural elements are clamped. From this point of view, the molecular mobility of a chain segment can be realized only with the beginning of the melting of the lamellas.

Thus, the specific nature of the manifestation of relaxation transitions in semi-crystalline polymers is due to the wide distribution of the kinetic elements responsible for these processes along the length, which in turn determines a wide range of their conformational states. As a result, in the disordered regions of polymers of this class, both β- and α-transitions of the same nature as in amorphous polymers and high-temperature transitions are realized, the parameters of which are determined by the presence of a crystalline phase.

The presented picture is applicable to the discussion of the relaxation behaviour of semi-crystalline polymers with a stable structure since it does not take into account the structural relaxation of the material during the test. The main regularities of these structural relaxation effects are discussed in Chapter 3.

3

The role of relaxation processes in phase transformations of polymers

The phase transitions of polymers include, first of all, phase transitions of the first kind – crystallization and melting under which the first derivatives of the Gibbs energy – enthalpy, entropy and volume – change abruptly.

Crystallization of polymers is determined by the integration of chain segments into a three-dimensional crystal lattice characterized by a high packing density. The possibility of the occurrence of such processes is determined by the chemical structure of the polymer, as well as by the configuration and conformational isomerism.

Only regular polymers with a clearly defined cofiguration, for example, iso- and syndtiotactic isomers, form ordered crystalline structures. Heterotactic polymers are not crystallized under any conditions.

For embedding in a crystalline lattice, fragments of macromolecules must assume a certain conformation, which ensures the greatest ordering and realization of the maximum possible packing density. For example, crystallization with the formation of lamellae proceeds due to the folding of macromolecules. The formation of such folds requires a high flexibility of the chain and a wide range of allowed conformations.

Bulk side substituents make dense packing of macromolecules difficult, thereby preventing crystallization of the polymer. On the contrary, the presence of polar atomic groups promotes crystallization due to the formation of strong intermolecular interactions that stabilize the crystalline phase that forms.

In accordance with these criteria, polymers are divided into crystallizable and non-crystallizing.

To clarify the role of relaxation phenomena during phase transformations of polymers, let us first consider the general thermodynamic and kinetic aspects of the crystallization and melting of this class of compounds.

From the point of view of thermodynamics at constant pressure, the phase transition proceeds with a decrease in the Gibbs energy

$$\Delta G = \Delta H - T\Delta S < 0, \tag{3.1}$$

and during melting, both enthalpy and entropy increase: $\Delta H_m > 0$, $\Delta S_m > 0$, and during crystallization the change in these thermodynamic functions has the opposite sign: $\Delta H_{cr} < 0$, $\Delta S_{cr} < 0$.

We write expression (3.1) for the process of polymer crystallization:

$$\Delta G_{cr} = \Delta H_{cr} - T\Delta S_{cr} < 0. \tag{3.2}$$

At the melting point $\Delta G_{cr} = 0$.

Consequently,

$$\Delta H_{cr} = T_m \Delta S_{cr}$$

or

$$T_m = \frac{\Delta H_{cr}}{\Delta S_{cr}}. \tag{3.3}$$

Taking into account expression (3.3), it is obvious that inequality (3.2) is satisfied at temperatures below the melting point, i.e. crystallization takes place at temperatures below T_m, or, in other words, the crystallization temperature T_{cr} is always lower than T_m.

For practical needs, consider the difference in these temperatures $\Delta T = T_m - T_{cr}$, which is referred to as the degree of overcooling. At $\Delta T \to 0$, the degree of crystallinity of the polymer (mass or volume fraction of the crystalline phase) and the crystallite size increase, and the defectiveness of the crystalline phase decreases. To obtain the most perfect crystalline polymers, crystallization is carried out at temperatures as close as possible to the melting point.

In the general case, the Gibbs energy of the crystalline phase is written as

$$G_{cr} = G_{cr}^{\infty} + \sigma s + \gamma, \tag{3.4}$$

where G_{cr}^{∞} is the Gibbs energy of an ideal crystal, σ is surface energy, s is the surface area, γ is the mole fraction of defects.

It follows from expression (3.4) that, other things being equal, the crystalline phase is thermodynamically the more favourable the smaller the specific surface of the crystals, i.e. the larger their size. An ideal is the infinitely large, defect-free single crystal.

We express the change in the Gibbs energy upon melting of the real crystalline phase:

$$\Delta G_m = G_{me} - G_{cr} = G_{me} - G_{cr}^{\infty} - \sigma s - \gamma = \Delta G_m^{\infty} - $$
$$- \sigma s - \gamma = \Delta H_m^{\infty} - T \Delta S_m^{\infty} - \sigma s - \gamma,$$

where G_{me} is the Gibbs energy of the polymer melt; ΔG_m^{∞}, ΔH_m^{∞}, ΔS_m^{∞} are changes in Gibbs energy, enthalpy and entropy upon melting of an ideal crystal of infinitely large dimensions.

At the melting point $\Delta G_m = 0$, hence

$$\Delta H_m^{\infty} - T_m \Delta S_m^{\infty} - \sigma s - \gamma = 0 \tag{3.5}$$

and

$$T_m = \frac{\Delta H_m^{\infty} - \sigma s - \gamma}{\Delta S_m^{\infty}} = T_m^{\infty} - \frac{\sigma s + \gamma}{\Delta S_m^{\infty}} = T_m^{\infty}\left(1 - \frac{\sigma s + \gamma}{\Delta H_m^{\infty}}\right), \tag{3.6}$$

where T_m^{∞} is the equilibrium melting temperature, i.e. the melting point of an ideal infinitely large single crystal: $T_m^{\infty} = \dfrac{\Delta H_m^{\infty}}{\Delta S_m^{\infty}}$.

Expression (3.6) indicates that the melting temperature of real crystals is below equilibrium, depending only on the energy of cohesion and, consequently, on the chemical structure of the substance. The observed difference is the greater the larger is the specific surface area of the crystal, i.e. specific surface energy, and the defectiveness of the crystalline phase.

Low-molecular substances form, as a rule, relatively large and perfect crystals, for which the contribution of surface effects and bulk defects is small. In connection with this, their melting is observed practically at a point with an abrupt change in the specific volume at a temperature that differs little from the equilibrium melting point (Fig. 3.1a).

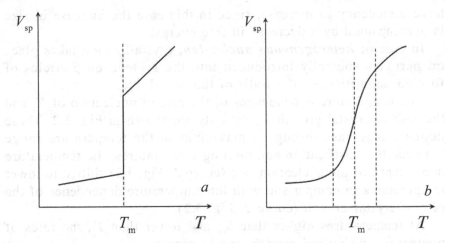

Fig. 3.1. The temperature dependence of the specific volume for a low-molecular-weight crystal (*a*) and a semi-crystalline polymer (*b*).

For a semi-crystalline polymer, melting is observed in the temperature range, the middle of which is considered as the melting point (Fig. 3.1*b*). This does not imply a violation of the thermodynamic requirement of a jump in the first-order phase transition. The melting of each individual crystallite occurs abruptly, however, semi-crystalline polymers are characterized by a wide distribution of crystallites in size and defectiveness, hence, at local melting points, which makes the transition smooth.

The reasons for the noted heterogeneity of crystallites in size, defectiveness, and melting points are related, first, to the non-uniform temperature distribution in the crystallizing melt – local degrees of overcooling differ in different microregions. Secondly, the formation of crystallites is influenced by the appearance of inner stresses that arise in the system when a viscous melt passes into a crystalline solid. Thirdly, the growth of crystallites formed at later stages is limited by neighbouring crystallites, resulting in steric hindrance, distortions of the crystal lattice, and the like.

From the point of view of kinetics, crystallization involves two stages – nucleation and growth of crystals. In this case, homogeneous and heterogeneous nucleation is distinguished.

In the case of ***homogeneous nucleation*** when the melt is cooled below T_m, spontaneous formation of fluctuation nuclei of the crystalline phase takes place. The free energy of the nucleus depends on its size and passes through a maximum at some critical value of this parameter. Only nuclei with a size exceeding the critical one have a tendency to increase, since in this case the increase in size is accompanied by a decrease in free energy.

In case of **heterogeneous nucleation**, crystallization takes place on particles specially introduced into the system, on particles of foreign impurities, on the walls of the vessel, etc.

The temperature dependences of the rate of nucleation of V_n and the rate of crystal growth V_{cr} crystals are shown in Fig. 3.2. These dependences pass through a maximum in the temperature range between the vitrification and melting temperatures, the temperature dependence of the nucleation rate (curve *1*, Fig. 3.2) shifted to lower temperatures in comparison with the temperature dependence of the rate of crystal growth (curve *2*, Fig. 3.2).

At temperatures higher than T_m and lower than T_c, the rates of nucleation and crystal growth tend to zero.

The appearance of a maximum on the temperature dependences of the rate of nucleation and crystal growth is explained by the following factors. Crystallization is thermodynamically 'allowed' at temperatures below T_m. As the crystallization temperature decreases, the thermodynamic driving force of this process increases and the rate of crystallization increases. However, crystallization is a diffusion-controlled process, since it involves the migration of segments from the melt to the surface of the growing crystal. The decrease in temperature is accompanied by an increase in the viscosity of the system, the diffusion of the segments becomes more difficult,

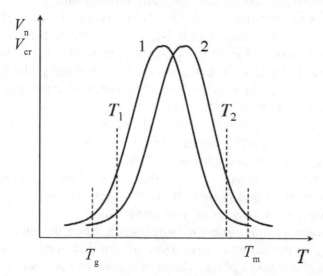

Fig. 3.2. The temperature dependences of the rate of nucleation (1) and crystal growth (2).

and the rate of crystallization decreases. The competition of these factors determines the extreme temperature dependence of the speeds of these stages of crystallization. At $T < T_g$, the crystallization is completely kinetically suppressed.

Thus, the results shown in Fig. 3.2, allow us to conclude that the ***crystallization of the polymer proceeds in the temperature range from the glass transition temperature T_g to the melting point T_m***. Similar trends are also observed in the crystallization of low-molecular-weight substances.

The kinetic regularities considered allow controlled formation of structural parameters of a semi-crystalline polymer and, as a result, a complex of operational properties of a final material.

Methodically, crystallization is carried out using two modes

1. isothermal crystallization at a given temperature;

2. crystallization upon cooling of the melt at a constant rate dT/dt.

In ***isothermal crystallization,*** the polymer melt is cooled to the selected temperature and maintained at a given temperature for a certain time. In this case, the parameters of the crystal structure of the material are determined by the temperature and the crystallization time.

During crystallization near the glass transition temperature (temperature T_1, Fig. 3.2), the nucleation rate is higher than the rate of crystal growth. As a result, a large number of nuclei appear in the system the growth rate of which is small. In this case, a fine-crystalline sample is obtained, characterized by low values of the modulus of elasticity and strength, but high deformability and plasticity. When crystallized near the melting point (temperature T_2, Fig. 3.2), the situation is exactly the opposite. The rate of growth of crystals exceeds the rate of nucleation, which leads to the formation of a large-crystal sample with high elastic modulus and strength, but low plasticity.

When investigating the kinetics of isothermal crystallization, crystallization isotherms are constructed – the dependence of the degree of crystallinity χ on time. The degree of crystallinity, i.e. mass or volume fraction of the crystalline phase, is estimated from the change in any experimentally measured parameter, for example, by volume reduction using the dilatometry method.

Typical crystallization isotherms are shown in Fig. 3.3. Their appearance is determined by the type of nucleation. In the case of heterogeneous nucleation, crystal growth begins almost

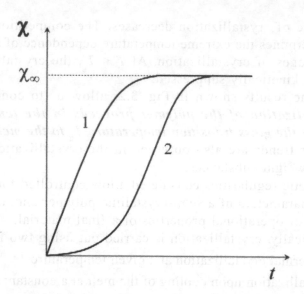

Fig. 3.3. Crystallization isotherms in the case of heterogeneous (1) and homogeneous (2) nucleation.

instantaneously (Fig. 3.3, curve *1*). For homogeneous nucleation, alag time is observed during which nuclei of critical size are formed, which can cause crystallization of the polymer (Fig. 3.3, curve *2*).

Processing of the experimental crystallization isotherms is based on the Kolmogorov–Avrami equation derived for low-molecular-weight substances. For semi-crystalline polymers, this equation is written in the form

$$\chi = \chi_\infty \left(1 - e^{-kt^n} \right), \tag{3.7}$$

where χ and χ_∞ are the current and the maximum degree of crystallinity attainable at a given temperature, k is the effective rate constant of crystallization, and n is the parameter that depends on the mechanism of nucleation and the shape of growing crystals.

Values of n close to integer are characteristic of homogeneous nucleation.

Fractional values are often obtained, which indicates the predominant role of heterogeneous nucleation (Table 3.1).

It should be taken into account that the Kolmogorov–Avrami equation was derived without taking into account the specific features of the crystallization of polymeric macromolecules. Therefore, the experimentally determined parameters k and n are usually considered as semi-empirical characteristics of polymeric crystallization

Table 3.1. The values of the parameter n of the Kolmogorov–Avrami equation

The regularity of crystal growth	Mechanism of nucleation	
	homogeneous	heterogeneous
One-dimensional (rods)	$n = 2$	$n = 1 - 2$
Two-dimensional (disks, plates)	$n = 3$	$n = 2 - 3$
Three-dimensional (spheres)	$n = 4$	$n = 3 - 4$

isotherms. For a more correct description of the mechanism of nucleation, the nature of crystallization, and the shape of growing crystals, independent structural studies are needed.

We also note that a distinctive feature of the kinetics of crystallization of polymers is the high sensitivity of the rate of crystallization to a change in temperature. This is especially true for small degrees of overcooling, i.e. at crystallization near the melting point. When the crystallization temperature is increased by several degrees, the rate of crystallization decreases by several orders of magnitude.

During crystallization upon cooling of the melt at a constant rate dT/dt = const, crystallization begins at a temperature below the melting point of the polymer and finishes at a temperature close to the glass transition temperature. It is obvious that as the cooling rate increases, the degree of crystallinity and the size of the crystallites decrease, and the defectiveness of the crystalline phase increases.

The thermodynamic and kinetic aspects of the phase transformations of polymers discussed above indicate that the characteristics of their crystal structure (the degree of crystallinity, the size of crystallites, the distribution of crystallites in size and defectiveness), and the melting point of the material depend to a large extent on the temperature–time regimes of crystallization and melting. For polymers, these processes have a pronounced relaxational, i.e. dependent on time nature.

The kinetic unit responsible for the crystallization and melting of the polymer is the segment of the macromolecule, and the elementary act of crystallization and melting is the transition of the segment from the melt to the crystal and vice versa. For polymers, the rate of such transitions is small and, as a rule, commensurate with the rate of change in temperature. The resulting relaxation effects can be demonstrated as follows.

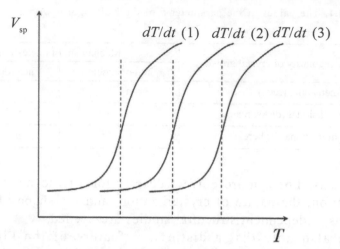

Fig. 3.4. The temperature dependences of the specific volume of the semi-crystalline polymer obtained at different heating rates $dT/dt(1) < dT/dt(2) < dT/dt(3)$.

Figure 3.4 shows the dilatometric curves of a semi-crystalline polymer obtained at different heating rates dT/dt.

An increase in the heating rate is accompanied by a shift of the curves to higher temperatures and an increase in the melting temperature, which is experimentally determined by the bending of the given dependences. A similar behaviour is characteristic of other methods for determining the melting temperature of polymers in the heating regime at a constant rate. For example, for differential scanning calorimetry, when the experimental values of melting points are fixed on the abscissa of endothermic melting peaks (Fig. 3.5).

Thus, unlike low-molecular-weight crystals for semi-crystalline polymers, the experimentally determined melting point is not a strict characteristic, since this parameter depends on the method and mode of measurement, primarily on the heating rate.

When analyzing the thermophysical behaviour during melting of semi-crystalline polymers the following parameters are determined:

- the experimental melting point T_m obtained by this or that method at a given heating rate (Figs. 3.4 and 3.5);
- the true (physical) melting point T_m^* of a given crystal structure of a given sample with a certain prehistory;
- the equilibrium melting point of a perfect crystal of an infinitely large size T_m^∞ (expression (3.6)).

Of these, only T_m^∞ is a fundamental characteristic, which is determined only by the cohesion energy and, consequently, by the chemical

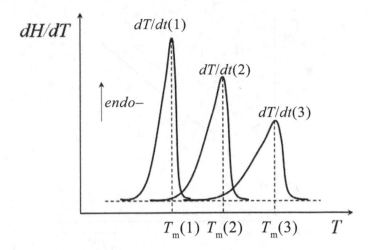

Fig. 3.5. Typical DSC melting curves of a semi-crystalline polymer, obtained at different heating rates $dT/dt(1) < dT/dt(2) < dT/dt(3)$.

structure of the polymer and does not depend on the crystallization conditions and test regimes.

Melting of lamellar crystallites of finite thickness l is observed at $T_m^* < T_m^\infty$. The ratio of these two temperatures is given by the Thomson–Gibbs equation:

$$T_m^* = T_m^\infty \left(1 - \frac{2\sigma}{\Delta H_m^\infty \rho l} \right), \qquad (3.8)$$

where σ is the free surface energy of the end faces, and ρ is the density of the crystal.

For a series of crystalline polymers with different values of l, determined by independent X-ray diffraction methods, expression (3.8) allows us to estimate the values of T_m^∞ by linear extrapolation of the dependence $T_m^* = f(1/l)$ to $1/l = 0$.

As noted above, the experimental melting point T_m is largely dependent on the heating rate of the sample (Figs. 3.4 and 3.5) and, in general, can increase by 15–20 degrees with increasing dT/dt by an order of magnitude.

The results of long-term studies of this effect by differential scanning calorimetry [20, 78–80] testify to the complex character of this dependence, the typical form of which is shown in Fig. 3.6.

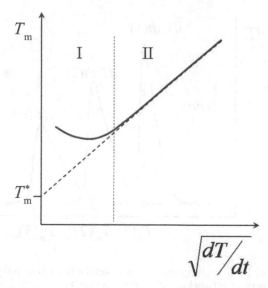

Fig. 3.6. A typical dependence of the experimental melting point of a semi-crystalline polymer on the heating rate.

In region II, usually corresponding to heating rates of 5 to 50 deg/min, the function $T_m = f\left(\sqrt{dT/dt}\right)$ is linear. From the point of view of molecular–kinetic concepts, the growth of T_m with increasing heating rate is explained by the fact that the segments, i.e. the kinetic units responsible for the melting of the polymer, do not have time to pass from the crystalline phase to the melt and, as a result, the melting of the material 'lags' behind the temperature rise. The higher the heating rate, the stronger this 'lag', and the melting of the polymer is experimentally fixed at an ever higher temperature. For low-molecular-weight crystals, the times of elementary melting events are orders of magnitude smaller than for polymers, so the change in the heating rate does not affect the value of the melting point.

The presence of a linear section on the dependence $T_m = f\left(\sqrt{dT/dt}\right)$ (region II, Fig.3.6) makes it possible to estimate T_m^* of the given crystal structure by extrapolating the experimental values to $\sqrt{dT/dt} = 0$, thereby excluding the influence of the heating rate on this characteristic. We note that the change in the molecular weight of the polymer in the range from 10^4 to 10^6 does not appreciably affect the value of T_m^*, which is determined only by the thermal prehistory of the samples and by the crystallization conditions.

In region I (Fig. 3.6) at low heating rates, i.e. at long times of thermal action, the sample has time to undergo structural relaxation associated with the improvement of metastable crystalline formations, recrystallization, increase in the size of crystallites, a decrease in their defectiveness, and so on. These processes lead to an increase in the melting point of crystallites, and the experimentally fixed T_m is a characteristic of a sample that is not original but modified during the test.

The relaxation nature of phase transitions in polymers is also manifested in the discrepancy between the melting point and crystallization temperatures when the sample is heated and cooled (Fig. 3.7).

When heated at a constant rate, the melting 'lags' behind the temperature rise, and the effective melting point is shifted to the right along the temperature scale. Upon cooling, the crystallization also 'lags', and the crystallization temperature shifts to lower temperatures. As a result, a pronounced hysteresis is observed.

For low-molecular-weight bodies, the direct (melting) and inverse (crystallization) processes pass through the same intermediate states, as a result of which these phase transformations jump abruptly at the same temperature.

Taking into account the relaxation nature of phase transformations of polymers allows to control the parameters of the crystal structure

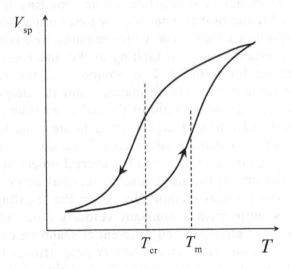

Fig. 3.7. Temperature dependences of the specific volume of a semi-crystalline polymer upon heating and cooling.

of the material, its melting temperature and, as a result, the final performance characteristics by **annealing** and **quenching.**

During **annealing**, the sample of the crystalline polymer is held for a long time (usually several hours) at temperatures close to the melting point. Partial melting of small and defective crystallites is observed, followed by recrystallization and the formation of larger and more perfect crystals.

Quenching involves the instantaneous cooling of the polymer melt to $T < T_g$. For kinetic reasons, the polymer does not have time to form a crystalline structure and is glassy, i.e. passes an amorphous vitreous state. The upper temperature of operation of such amorphized materials is their glass transition temperature T_g. Above T_c segmental mobility appears in the polymer, which is accompanied by spontaneous crystallization of the sample.

In metallurgy, quenching of alloys of a certain composition with cooling at rates of $\sim 10^6$ deg/s is used to produce metallic glasses possessing a unique combination of mechanical, magnetic, electrical and anticorrosive properties.

Thus, in polymers, first-order phase transitions (crystallization and melting) have a pronounced relaxation character. This is manifested, first of all, in the dependence of the experimental melting temperature on the heating rate of the sample (Figs. 3.4 and 3.5) and the melt and crystallization mismatch in the 'heating–cooling' cycle (Fig. 3.7). The basis of this behaviour is the chain nature of macromolecules.

The elementary act of crystallization and melting involves the transfer of the kinetic unit responsible for these processes, from the melt to the crystal and vice versa, which requires a certain time τ. In practice, the crystallization and melting of the material is provided by thermal action for a time t. It is obvious that the very fact of the course of these phase transformations and the degree of their completeness depends on the ratio of the indicated time parameters.

For low-molecular-weight crystallizing bodies, the kinetic unit that determines the elementary act of crystallization and melting is a molecule or atom, for which the time τ is several orders of magnitude smaller than the practically used time of thermal action t ($\tau \ll t$). For modern experimental methods based on the 'heating–cooling' regimes of a sample with a constant velocity (for example, the above-mentioned dilatometry and differential scanning calorimetry), the available variation dT/dt and inversely proportional to time t is 2–3 orders of magnitude. Variation of the test time interval in this interval does not violate the relation $\tau \ll t$, and as a result, low-molecular-weight bodies melt and crystallize at one strictly defined temperature, independent of the experimental conditions.

For polymers, phase transitions are determined by the transfer of segments for which the value of τ is comparable to the value of t ($\tau \sim t$). In this case, even a slight (several times) change in the rate of thermal action leads to the appearance of the relaxation effects noted above.

We note in particular that the observed effects are clearly manifested also in the structural relaxation of crystallized polymers.

The most perfect, coarse-crystalline structure can be achieved when the polymer is crystallized from dilute solutions. During crystallization from a melt, a crystalline sample is characterized by the presence of a large number of metastable crystallites with a wide distribution in size and defectiveness. Subsequent annealing of such samples at elevated temperatures approaching the melting interval is accompanied by a structural reorganization of the material due to various processes, among which, first of all, we note

- pre-crystallization of the amorphous component of the semi-crystalline polymer;
- recrystallization, including partial or complete melting of unstable initial crystallites with further formation of larger and perfect crystalline formations;
- polymorphic transformations with the formation of more stable crystallographic modifications;
- secondary phase transitions preceding melting.

The listed variants of structural relaxation are also caused by the transfer of kinetic units (segments), which requires a certain time τ. It is obvious that, as in the case of macroscopic crystallization and melting, the possibility of the occurrence of such microphase transformations is dictated by the ratio of the time τ and the time of the thermal action t.

Consider the patterns of this behaviour using the differential scanning calorimetry method [20].

In this technique, the time of thermal action t is given by the heating rate of the sample. If the heating rate is much greater than the rate of structural relaxation ($t \ll \tau$), during the test the crystal reorganization does not have time to take place. As a result, the DSC curve is characterized by a single endothermic peak corresponding to the melting of the original metastable structure (curve *1*, Fig. 3.8).

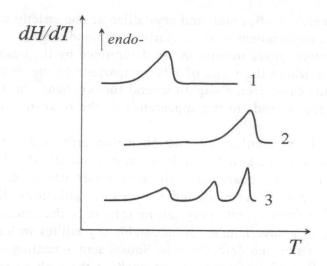

Fig. 3.8. Endothermic melting peaks of a semi-crystalline polymer at different heating rates. Explanations in the text.

At low heating rates the condition $t \gg \tau$ is fulfilled, which ensures complete structural relaxation of the initial structure. In this case, one endothermic peak is also observed on the DSC curve, corresponding to the melting of more perfect and large crystallites that appeared during the heating process (curve *2*, Fig. 3.8). Obviously, the melting temperature of such a 'relaxed' structure is higher than that of the original one.

When the heating rate is comparable with the rate of structural relaxation ($t \sim \tau$), multiplet melting endotherms appear on the DSC curves (curve *3*, Fig. 3.8). This experimental result reflects the sequence of acts of partial or complete melting of primary crystallites, their reorganization into more stable formations, melting of the emerging crystal structure, etc. We note that in a number of cases it is possible to fix exothermic peaks of crystallization between the melting endotherms.

From the practical point of view, as mentioned above, the relaxation character of phase transformations of polymers allows the controlled structural modification of crystallizable polymers from a completely amorphized state to the formation of coarse crystalline, stable crystal structures by selecting special temperature-time modes of action. This, in turn, determines a wide range of the physico-mechanical and functional properties of this class of polymeric materials.

4

Relaxation aspects of plastic deformation of polymers

Plastic deformation of polymers is a complex structural and mechanical process the comprehensive aspects of which have been the subject of many years of research [4,20,21,81–86].

The interest in this problem is obvious – the fundamental work on the study of plastic deformation is the basis for the development of scientific and technological principles for the creation of modern plastics characterized by a unique combination of mechanical properties, namely elasticity, strength and impact strength. This determines the use of this class of polymer materials in various fields of technology, including machinery and shipbuilding, aerospace, construction, etc.

Modern plastics are based on two classes of polymers. The first class includes polymeric glasses, i.e. completely amorphous polymers with a glass transition temperature T_g lying in the range from 100°C to 400°C and much higher than their service temperature T_{ser}. The second class includes semi-crystalline polymers that are used at temperatures below their melting point T_m.

To understand the nature of plasticity, a number of model representations and mechanisms of the deformation process are involved.

The mechanism of plastic deformation of semi-crystalline polymers is interpreted from the point of view of experimentally fixed structural rearrangements of their supramolecular, crystalline structure. These questions are discussed in more detail in Section 4.2.

For the amorphous polymeric glasses, direct diffraction methods for analyzing the initial structure and its evolution during deformation are not applicable, as a result the nature of the plastic deformation of these polymers is still a subject of debate.

The main approaches for describing the plasticity of glassy polymers can be divided into two groups.

The first of them is based on the discussion of the plastic deformation of polymeric glasses from the molecular–kinetic point of view, thoroughly developed to describe the viscous flow of liquids [8–10, 87]. We note that these views are successfully used when considering the high-elastic deformation of rubbers. The applicability of similar approaches to glasses (amorphous polymers below their glass transition temperatures) is based on treating glass as a supercooled liquid that is identical in structure to rubbers. From these positions, deformation mechanisms due to the segmental mobility of polymer chains, their unfolding during deformation, etc. can be applied to plastic (forced-elastic) deformation of glassy polymers. Let us mention here the description of plastic deformation and the viscous flow of both polymeric and non-polymeric glasses in the framework of the model of excited atoms [88–91]. This model is based on the concept of the dominant role of the critical displacement of atoms or a group of atoms relative to the equilibrium state in molecular–kinetic processes. Such critical displacements of kinetic particles with their local rearrangements are to some extent taken into account in many theories of molecular mobility in high-viscosity liquids and glasses.

The second group of approaches for describing the plastic deformation of glassy polymers is based on the provisions of the general theory of plasticity of metallic and crystalline bodies [86, 92–106]. The use of such approaches is justified by the fact that the macroscopic picture of the plastic deformation of vitreous and crystalline bodies has extremely many similarities, namely the presence of a yield point, the localized nature of deformation, thermophysical effects, etc.

A brief review of the existing theories concerning the plastic deformation of polymeric glasses is given in [15]. Here we also discuss some relaxation aspects of the plastic deformation of polymers of various classes that determine the regularities of this process.

4.1. Polymeric glass

First of all, let us consider the behaviour of polymeric glasses under deformation conditions at a constant rate. Methodically, this experiment is usually carried out in uniaxial tension, uniaxial compression and shear modes. The response of the polymer is fixed in the form of a deformation curve, i.e. the dependence of the stress σ on strain ε ($\sigma - \varepsilon$ diagram), a typical form of which is shown in Fig. 4.1.

Three sections are identified on this curve. Section I is bounded by the yield point with the coordinates σ_y and ε_y. The slope of the initial linear stress–strain relation is the modulus of elasticity or Young's modulus E_0. In the transition from section I to section II, the deformation curve passes through a maximum (yield tooth), after which a steady-state stage of deformation is observed within the limits of section II at a constant stress. In section III, the increase in deformation proceeds with a noticeable increase in stress.

We note that the above-mentioned theories of the plasticity of glassy polymers refer to the description of the deformation mechanism at the initial stage of the process (section I). It is generally accepted that the steady-state development of deformation in Section II is determined by the unfolding of macromolecular coils due to segmental mobility, activated by the combined effect of thermal and force fields (the mechanism of yielding). At this temperature, the

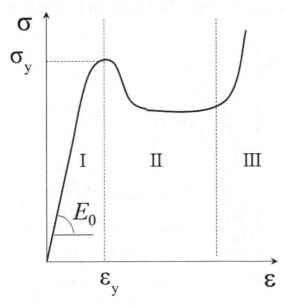

Fig. 4.1. A typical stress–strain diagram corresponding to the plastic deformation of a polymeric glass.

conditions for similar molecular–kinetic rearrangements are realized when the stress σ_y is reached (Fig. 4.1).

These microscopic processes are localized in certain regions of the sample – the 'neck' in the uniaxial tension and shear bands, which arise at an angle of 45° towards the axis of deformation under shear. The results of direct observations of the localized nature of the plastic deformation of polymeric glasses can be found in a number of works, for example, Refs. [86,107–112].

Section III of the deformation curve is designated as the stage of orientational hardening, within which the load is placed on oriented macromolecular chains or their fragments.

The relaxation nature of the plastic deformation of polymeric glasses is clearly manifested in the dependence of both the profile of the deformation curve and the mechanical parameters (Fig. 4.1) on the temperature and time conditions of deformation.

To understand the relaxation effects that determine the development of deformation, consider the following test regimes:

1. change in the deformation temperature T_{def} at a constant strain rate $V = d\varepsilon/dt$;

2. change in V for a constant T_{def}.

The effect of temperature on the deformation behaviour of glassy polymers at a constant rate of deformation is shown in Fig. 4.2.

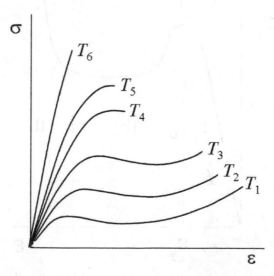

Fig. 4.2. The ²strain–strain curves of polymeric glass at temperatures $T_1 > T_2 > T_3 > T_4 > T_5 > T_6$ and constant strain rate.

A decrease in the temperature in the range from T_1 to T_3 leads to a shift of the deformation curves to the region of higher stresses and an increase in the values of $\sigma_{f.e.}$, $\varepsilon_{f.e.}$ and E_0. A further decrease in temperature is accompanied by a degeneration of the stage of steady-state development of deformation, loss of plasticity and transition at temperature T_6 to brittle fracture.

The nature of the 'brittleness–plasticity' transition in the temperature range from T_3 to T_6 is considered in Chapter 6. Here we shall confine ourselves to a discussion of the effect of temperature on the regularities of plastic deformation of glassy polymers (temperatures T_1–T_3, Fig. 4.2).

In the framework of the mechanism of yielding when the stress σ_y is reached the segmental mobility is activated in the polymeric glass. The translational movements of the segments, determinine deformation or unfolding of macromolecular coils. Such translational motion of kinetic units requires overcoming the activation barrier E_a (see Section 1.2). This becomes possible when, under the experimental conditions, the sum of thermal kT and mechanical energy $\gamma\sigma$ becomes larger than the activation barrier energy ($kT + \gamma\sigma > E_a$).

At a given temperature, i.e. for a given contribution of kT, this requirement is satisfied in the yield point at a stress of σ_y. It is obvious that as the temperature increases, the contribution of kT increases, the activation of segmental mobility is observed at lower values of σ_y, and plastic deformation occurs at lower stresses. These molecular–kinetic considerations explain the evolution of the deformation curves with a change in temperature observed in practice (Fig. 4.2).

At a constant temperature, the influence of the strain rate on the course of deformation curves is reflected in Fig. 4.3.

We note, first of all, the fact that the change in the strain rate at a given temperature is opposite to a change in the deformation temperature for a given rate. In other words, a decrease in the strain rate affects the deformation behaviour of the polymer in a manner similar to an increase in temperature, and vice versa.

This behaviour can be described in terms of temperature–time superposition (see Section 1.4) – under deformation conditions, the same polymer response or the same 'stress–strain' curve can be obtained by selecting a specific temperature for a given strain rate or, conversely, by selection of the strain rate at a given temperature. To understand the relaxation nature of the observed effects, let us

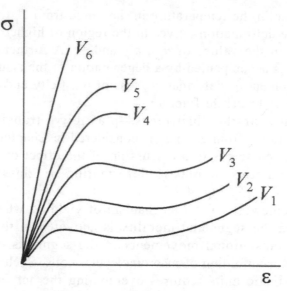

Fig. 4.3. The 'stress–strain' diagrams of polymeric glass obtained at strain rates $V_1 < V_2 < V_3 < V_4 < V_5 < V_6$ and constant temperature.

turn to the molecular–kinetic picture of plastic deformation (see Section 1.2).

Thus, the elementary act of plastic deformation upon reaching the yield stress σ_y is the translational movement of a segment from one potential well to another with the overcoming of the activation barrier. This event requires a certain time τ, which decreases with increasing temperature. The strain rate V determines the exposure time t, which decreases with increasing V.

Plastic deformation due to the above elementary transitions of segments easily develops, when $t \gg \tau$, i.e. at relatively low strain rates or elevated temperatures. As the strain rate increases or the temperature decreases $t \to \tau$, the segments do not have time to regroup, the delay of microscopic rearrangements with respect to macroscopic action is observed, and plastic deformation is realized at higher values of $\sigma_{f.e.}$. Further increase in the strain rate or a decrease in temperature leads to the fact that t becomes much smaller than τ ($t \ll \tau$), which means a kinetic inhibition of the plasticity of the material and the transition to brittle fracture.

Thus, the plastic deformation of the polymeric glass has a pronounced relaxation character – the quantitative characteristics and the profile of the 'stress–strain' curve, as well as the mechanical parameters of the material (first of all, the stress and strain corresponding to the yield point σ_y and ε_y, respectively) are determined by the temperature–rate regimes of mechanical action. Therefore, in certification practice, the tabulated mechanical parameter of the polymer is measured strictly under the conditions specified by the relevant standards.

At the same time, the fundamental question remains about the possibility of a universal description of the strain curve of a polymeric glass, free from the influence of the temperature–time conditions of the experiment.

Attempts at a unified analysis of the physico-mechanical behaviour of polymers and materials based on them were undertaken in [113, 114] when considering their mechanical properties in the corresponding states. To generalize the results of mechanical tests, the principle of temperature–time superposition [13], whose applicability for the analysis of the mechanical behaviour of amorphous and crystalline polymers was discussed in [115–118], is also widely used. A number of generalized relationships describing the physico-mechanical behaviour of polymeric glasses are proposed in Refs. [119,120].

In the cycle of papers [15,60,114,115,121–126] the procedure for unifying 'stress–strain' curves of glassy polymers has been proposed and experimentally substantiated. Attention was mainly paid to Section I (Fig. 4.1) with deformations not exceeding ε. Interest in this region of the σ–ε diagram is due to the fact that, from consumer positions, the isotropic glassy polymer can only be used in this strain range. As noted above, when the yield point is reached, strain is localized in the 'neck' or shear bands develop in the polymer. From the operational point of view, this means the loss of material integrity and instability of its mechanical behaviour.

It was shown that in the deformation region $\varepsilon \leq \varepsilon_y$ unification the deformation curve $\sigma = f(\varepsilon)$ is achieved by normalizing the current values of stress σ and deformation ε by the values of σ_y and ε_y, respectively.

The deformation curve, reconstructed in the dimensionless, reduced coordinates $\dfrac{\sigma}{\sigma_y} = f\left(\dfrac{\varepsilon}{\varepsilon_y}\right)$, (Fig. 4.4) is a stable characteristic of the process of plastic deformation, the profile and quantitative parameters of which do not depend on the chemical structure of the polymer, the conditions and the test regime.

Thus, the proposed procedure for processing the experimental deformation curves in the reduced, dimensionless coordinates neutralizes the relaxation nature of plastic deformation and provides a

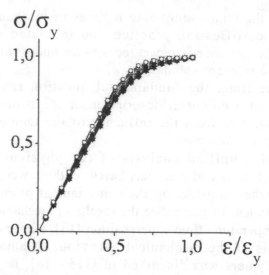

Fig. 4.4. Unifed 'stress–strain' curve of polymeric glass.

description of the process free from the influence of the deformation conditions. The essence of the result obtained is that the apparent relaxation effects are 'hidden' in the normalizing parameters of the σ_y and ε_y.

The details and nature of this methodology are set out in Appendix 4. Here we only note that the above approach for unifying deformation behaviour is not specific for polymeric glasses. Analysis of the data [127–131] indicates that the universal description of the deformation behaviour in terms of the generalized deformation curve (Fig. 4.4) is applicable to low-molecular-weight organic, inorganic and metallic glasses, as well as to a number of metallic materials. The exception is formed by crystallizing polymers, as well as complex alloys, mixtures and composites that undergo phase and polymorphic transformations during deformation.

It is noteworthy that computer modelling of deformation for a planar disk system associated with non-valent interactions [132] leads to the construction of a deformation curve that completely satisfies the experimentally obtained unified dependence (Fig. 4.4). This result allows us to conclude that the above version of the unification of plastic deformation extends not only to certain classes of materials (polymeric and metallic glasses, but also metals), and can be used for the universal description of plastic systems as a whole.

4.2. Semi-crystalline polymers

The range of the currently used semi-crystalline polymers is extremely wide and includes polyolefins (primarily polyethylene and polypropylene), aliphatic and aromatic polyamides, polyesters and polyethers, etc. Details of the deformation behaviour of such polymeric materials are escribed in a huge number of original works. In this section, we consider the most general regularities of plastic deformation of this class of polymers [4,83–85,133,134].

The deformation of semi-crystalline polymers is accompanied by experimentally fixed rearrangements of the initial crystal structure of the material[1], as well as by phase and polymorphic transformations. For specifics, we will analyze the structural and relaxation pattern of plastic deformation of films and fibers based on semi-crystalline polymers in the uniaxial tension regime.

The macroscopic picture of the deformation of these materials resembles that of polymeric glasses (see Section 4.1). For example, when a film is stretched (the axis of deformation is shown by arrows, Fig. 4.5) in the yield tooth, i.e. at the transition from section I to section II, the formation of a 'neck' occurs – a sharp local narrowing of the sample, which is discernible to the naked eye. For clarity, in Fig. 4.5 the 'neck' is highlighted in black. At the stage of steady-state stageof deformation (section II), the 'neck' gradually grows along the length of the sample due to the reduction of the undeformed part. In other words, the material of the undeformed part 'flows' into the 'neck', and this process is localized on the boundary between them.

The transition from section II to section III (to the stage of orientational hardening) is observed when the sample completely passes into the 'neck'.

At the microscopic level, the nature and mechanisms of plastic deformation of the semi-crystalline polymers and polymeric glasses are radically different from each other.

In section I, the deformation, the mechanical response of a semi-crystalline polymer is determined mainly by its amorphous phase, constructed from the tie-chains. The mobility of the tie-chains and, consequently, the mechanical behaviour of the amorphous phase is controlled by the quantitative parameters of the crystal structure – the degree of crystallinity and the size of the crystals. With an increase in these two characteristics, the mobility of the tie-chains decreases,

---

[1]The structure of semi-crystalline polymers is described in Appendix 3.

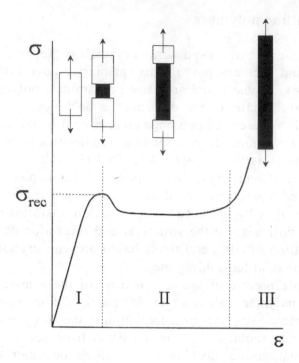

Fig. 4.5. A typical deformation curve of a semi-crystalline polymer. Explanations in the text.

which is accompanied by an increase in the rigidity of the sample and an increase in its modulus of elasticity.

When going from section I to section II, the material is subjected to an external stress-induced destruction of the initial spherulitic or lamellar structure and the appearance of a 'neck'. This process is associated with the crushing of the crystallites and the orientation of their fragments along the extension axis. Under the conditions of uniaxially applied deformation, fragments of destroyed spherulites or lamellas form a new fibrillar crystal structure that is characterized by a high degree of orientation of the material.

The primary structural element of the fibrillar structure (fibril) is an anisodiametric structural formation with a huge (several orders) ratio of length to thickness, constructed by alternating crystalline and amorphous regions (Fig. 4.6).

The crystalline fibril regions are constructed from macromolecular chains in a straightened or folded conformation. The same macromolecule participates in the formation of a number of crystallites both within a single fibril and several crystallites of neighbouring fibrils. Intra- and interfibrillary tie-chains form intra-

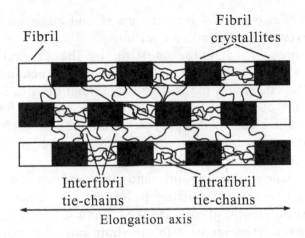

Fig. 4.6. Schematic representation of the oriented fibrillar structure of a semi-crystalline polymer.

and interfibrillar amorphous regions of the oriented structure of the polymer.

Within this fibril, a crystallite with an adjacent amorphous region forms a large period along the axis of the fibril. The linear size of a repeat period increases with an increase in the deformation temperature and during subsequent annealing of the oriented sample and ranges from several units to several tens of nanometers.

In parallel, the fibrils laid along the extension axis form an oriented 'neck' structure.

In the region of steady-state development of deformation (section II) after a peak on the stress–strain curve, deformation proceeds at a constant stress due to the growth of the 'neck' along the length of the sample. The constancy of the stress is due to the fact that the mechanically activated fibrillation of the polymer is localized at the boundary between the 'neck' and the isotropic material, and the stress required for this is independent of the magnitude of the deformation, i.e. the length of the 'neck'.

In Section III (the stage of orientational hardening), when the 'neck' extends over the entire length of the specimen, the material is completely transformed into an oriented state, the deformation of which requires an increased stress.

The considered mechanism of deformation of a semi-crystalline polymer is interpreted as follows.

Destruction of the original crystal structure can be described as mechanically activated crystallite melting. This process occurs under the action of two fields – temperature kT and mechanical $\gamma\sigma$, where σ is the external stress, γ is a coefficient.

Mechanically activated melting of the crystalline phase requires a certain amount of thermal and mechanical energy $kT + \gamma\sigma$. Together with the subsequent formation of a new fibrillar structure, the observed structural rearrangements are interpreted as the recrystallization of the polymer. The necessary stress for this temperature corresponds to a peak on the 'stress–strain' curve, which is called the recrystallization limit σ_{rec}.

The influence of temperature and strain rate on the deformation behaviour of semi-crystalline polymers is similar to that for amorphous polymeric glasses (Figs. 4.2 and 4.3) – with a decrease in temperature or an increase in the strain rate, the growth of σ_{rec} is observed and a transition to brittle fracture occurs. The regularities of the 'brittleness–plasticity' transition are considered in Chapter 6. Here we will briefly discuss the relaxation nature of the plastic deformation of semi-crystalline polymers, which is clearly traced in the temperature–rate dependences of the recrystallization stress.

Obviously, at a constant strain rate, a decrease in the deformation temperature means a decrease in the contribution of the thermal energy kT to the mechanically activated processes of melting and recrystallization of the polymer. This naturally requires a larger contribution of mechanical energy $\gamma\sigma$, as a result of which there is an increase in σ_{rec} and a vertical shift of the $\sigma-\varepsilon$ diagrams towards higher stresses.

The destruction of crystallites and the formation of a fibrillar structure proceeds due to the movement of kinetic units (segments of macromolecules) from one position to another, which requires a certain time τ, depending on the temperature. At a given temperature, an increase in the strain rate or, which is the same, a decrease in the time of action t, results in the displacement of the segments 'lagging' the deformation development. As a result, the effective limit of recrystallization σ_{rec} is observed to increase, which ultimately causes the plastic deformation of the semi-crystalline polymer to proceed at higher values of external stress σ.

A unified approach for describing the physico-mechanical behaviour of semi-crystalline polymers analogous to that developed for polymeric glasses (see Section 4.1) is not possible. The reason for this is that, unlike yielding of glassy polymers, recrystallization involves complex structural rearrangements directly during deformation. These changes determine the constantly changing mechanical response of the polymer to the external action. The nature of the occurrence of such structural–mechanical processes depends on the initial structure of a particular sample, its degree of crystallinity, the distribution of crystallites by size, and so on. Moreover, even at the section I of the 'stress–strain' curve, i.e. up to the recrystallization stress, structural transformations are observed that are associated with the orientation of the tie-chains and crystallites along the tensile loading axis, the crushing of fine crystallites, and the like.

5

Relaxation of deformed polymers

The study of the nature, mechanisms and regularities of the relaxation behaviour of polymers subjected to various mechanical influences is directed, first of all, to the solution of a number of practically important problems, among which we will outline the following.

Processing of polymers by orientational drawing, stamping, pressing, casting, etc. includes mechanical (including plastic) effects on the substance. As a result, residual strains and internal stresses that can relax over time at these 'working' temperatures accumulate in the final product.

Under the operating conditions, materials often experience alternating (cyclic) loads. In this case, the durability of the material and the stability of the product are determined by the depth of the course of the relaxation processes, i.e. completeness of 'dumping' when unloading strains and stresses accumulated during loading.

On the other hand, the study of relaxation processes taking place in a deformed material also has a pronounced fundamental aspect since it serves as a basis for the development of structural and mechanical models of the behaviour of materials and the theoretical mechanisms of their deformation.

In this chapter, an analysis is made of thermally stimulated relaxation phenomena in polymers subjected to various mechanical influences.

First of all, we note a fundamental difference in the effect of temperature on the behaviour of deformed samples of semi-crystalline polymers and polymer glasses.

Plastically deformed semi-crystalline polymers when heated to the melting temperature T_m (the upper temperature of operation of this class of materials) become liquid viscous and completely lose shape.

The flow of thermally stimulated relaxation processes in a plastically deformed material can be convenicntly monitored by the temperature dependences of the residual strain $\varepsilon_{res} = \dfrac{l_0 - l_T}{l_0 - l}$, where l_0 is the initial length of the sample, l is the length of the deformed sample, and l_T is the length of the deformed sample at temperature T.

As an example, this dependence is shown in Fig. 5.1 (curve *1*) for a uniaxially oriented PA-6 film.

As noted in Section 4.2, the orientation stretching of semi-crystalline polymers at temperatures much lower than the melting point T_m is accompanied by experimentally recorded structural rearrangements of the crystalline phase, namely mechanically activated 'melting' of the original crystalline (lamellar, spherulitic, etc.) structure and the formation of a new fibrillar, oriented structural modification.

The complex nature of thermally stimulated relaxation or 'shrinkage' of the oriented polymer at $T < T_m$ (Fig. 5.1, curve *1*) is due to relaxation processes in the amorphous phase of the polymer, as well as partial melting of the crystallites.

The observed relaxation can be effectively prevented by high-temperature annealing of the deformed sample in the stressed state. The material subjected to such a procedure almost completely retains the predetermined deformation to temperatures close to its T_m (Fig.

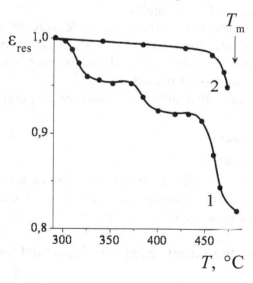

Fig. 5.1. The temperature dependences of the residual deformation ε_{res} for a uniaxially oriented film PA-6 (1) and a uniaxially oriented PA-6 film subjected to high-temperature annealing in a stressed state (2).

5.1, curve *2*). This technique is widely used in industry for the production of heat-resistant oriented films and fibers.

Plastically deformed amorphous polymeric glasses when heated to the glass transition temperature T_g (the upper operating temperature of this class of materials) take their original shape and dimensions. In other words, in this case there is no loss of material or product, but its return to the original undeformed state.

Thus, for polymeric glasses, plastic deformations are reversible – annealing of the deformed polymer at the glass transition temperature completely 'erases' the deformation history of the sample.

The nature of this behaviour lies in the fact that plastic (forced-elastic) deformation of the glassy polymer does not lead to an irreversible displacement of macromolecules as a whole (see Section 4.1). Macroscopic deformation is due only to the unfolding of macromolecular coils due to segmental mobility, activated by joint exposure to temperature and applied mechanical stress. When unloading, the microscopic deformation of macromolecular coils is fixed in the polymer, thereby determining the macroscopic residual deformation of the sample.

Increasing the temperature to T_g activates the segmental mobility, ensuring the return of the deformed polymer coils to the initial state, which is characterized by the maximum of entropy. Macroscopically this is expressed in the complete restoration of the original dimensions and shape of the sample.

Taking into account the noted unique feature of plastically deformed vitreous polymers associated with the restoration of the initial state upon heating, let us dwell in more detail on the regularities of this relaxation process.

First of all, we note that deformed samples of polymer glasses are characterized by

- stored latent energy;
- residual deformation.

When the temperature is raised, relaxation of both the energy and deformation parameters is observed. Let us consider the main regularities of these processes.

5.1. Relaxation of the latent energy of deformed polymer glasses

In addition to the obvious change in the size and shape of the sample the plastic deformation of the polymer glass is accompanied by an increase in the enthalpy of the material ΔH, equal to, assuming a constant volume ($p\Delta V \approx 0$), an increase in internal energy. These effects are quantitatively determined by differential scanning calorimetry (DSC) [20].

Figure 5.2 shows the typical DSC curves for the initial polymer glass (curve *1*) and the plastically deformed sample (curve *2*). The area between these curves makes it possible to estimate the excess enthalpy ΔH for a deformed sample. As mentioned above, this increase in enthalpy caused by deformation of the polymer is interpreted as stored or latent energy.

At present, there is no single view of the nature of the observed change in the energy state of the polymer.

These effects are associated with the following factors [20,62, 93,135–138]:

- change in intramolecular energy due to conformational *trans-gauche* transitions;
- change in the energy of intermolecular interactions;
- distortion of valence angles and bonds;
- accumulation of internal stresses;
- development of a new surface as a result of the formation of microcracks and crazes, as well as fibrillation of the material;
- the emergence and evolution of high-energy, small-scale shift transformations.

A critical review of the concepts and theories describing the issue

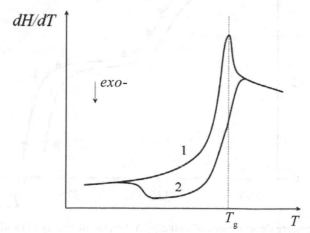

Fig. 5.2. Typical DSC curves for the original glassy polymer (1) and for the plastically deformed sample (2).

raised lies outside the scope of this monograph. We only note that, in our opinion, the most experimentally developed approach linking the accumulation of internal energy with a change in molecular packing and, accordingly, the energy of intermolecular interaction (IMI), as will be discussed below.

As follows from the data shown in Fig. 5.2, when the deformed sample is heated to temperatures in the vicinity of T_g, the stored energy completely disappears ($\Delta H \rightarrow 0$), and the polymer returns to the initial energy state. However, an appreciable fraction of ΔH relaxes at temperatures much lower than the glass transition temperature (Fig. 5.3, curve *1*).

A study of deformation and relaxation behaviour for a model series of polymeric glasses, styrene–methacrylic acid copolymers, was undertaken in Refs. 139–142.

For the samples studied, infrared spectroscopy data were used to evaluate

- parameters of the universal, van der Waals IMI, due to the interaction of benzene rings;
- parameters of a specific IMI due to hydrogen bonding between carboxyl groups of methacrylic acid (a system of single and double hydrogen bonds with an energy of 19 and 38 kJ/mol, respectively);

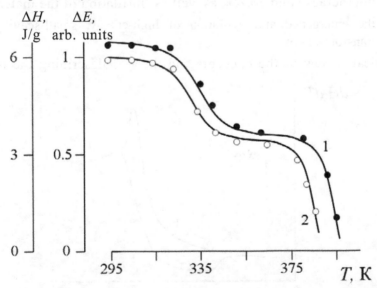

Fig. 5.3. Temperature dependences of enthalpy ΔH or latent energy (1) and hydrogen bonding system ΔE (2) for styrene copolymer with 16 mol. % methacrylic acid deformed in the uniaxial compression mode by 30% at 20°C [20].

- parameters of intramolecular energy, i.e. the difference between the energies of *trans-* and *gauche-*conformers.

The parameters of molecular mobility were determined from the areas of temperature dependences of the loss tangent for the initial and deformed samples.

The enthalpy of the deformed samples was evaluated by the differential scanning calorimetry method using the area of the exothermic effect caused by deformation (Fig. 5.2).

The obtained results showed that

1. with $\varepsilon < \varepsilon_y$ (see Fig. 4.1), the deformation does not have any effect on the parameters of the universal and specific IMI, the enthalpy, and also the molecular mobility in the polymer;

2. with $\varepsilon \geq \varepsilon_y$ there is a decrease in the parameters of the universal and specific IMI, the growth of the enthalpy and molecular mobility of the samples;

3. The parameters of intramolecular energy practically do not change in the entire range of deformations studied (up to 60%).

Thus, the increase in enthalpy as a result of deformation should be associated with a decrease in the energy of the IMI in the polymer glass, which also explains the growth of molecular mobility in deformed samples. At a qualitative level, this picture is the same for all types of loading used (uniaxial compression and tension, shear, rolling, torsion). Quantitative effects are maximum expressed for uniaxial compression and rolling.

We emphasize particularly that when the deformed polymer is heated, the reduction of the hydrogen bond system, disturbed by the preliminary deformation, proceeds in parallel with the enthalpy or latent energy relaxation (Fig. 5.3). The beginning of these relaxation processes is associated with the temperature of the relaxation β-transition T_β (see Section 2.1.1) [20]. In other words, the relaxation of the latent energy takes place in the temperature interval $T_\beta - T_\alpha$, completely ending at T_α.

For a complete understanding of the relaxation effects manifested in deformed polymer glasses, let us turn to the laws of thermostimulated recovery of their dimensions.

5.2. Relaxation of the dimensions of deformed polymer glasses

The main results given in this section relate to the relaxation of glassy polymers deformed in uniaxial compression mode [15,62,125, 143,144]. Control experiments showed that the picture of relaxation processes does not differ qualitatively for samples subjected to uniaxial stretching and shear.

The experimental procedure included

- uniaxial compression of cylindrical specimens to a certain value of deformation ε at a given temperature and strain rate;
- cooling the deformed sample in a loaded state with liquid nitrogen;
- unloading at liquid nitogen temperature; heating at a constant speed of 1°/min.

Relaxation was recorded by changing the linear size of the sample. The current deformation was estimated as $\varepsilon_T = \dfrac{h_0 - h_T}{h_0}$, where h_0 and h_T are the initial linear size of the sample and the linear dimension of the deformed sample at temperature T, respectively. The linear size of the sample was measured with an accuracy of ±0.01 mm.

Figure 5.4 shows on the example of poly(methyl methacrylate) (PMMA) a typical curve of thermally stimulated relaxation of the dimensions of a deformed polymeric glass.

Attention is drawn to the similarity of this process to the relaxation of latent energy (Figure 5.3). When the plastically deformed sample is heated, an appreciable part of the deformation is recovered at temperatures well below the temperature of the relaxation α-transition

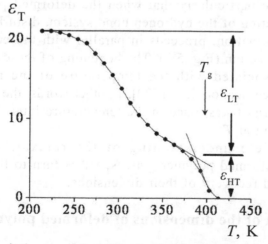

Fig. 5.4. A typical curve of thermally stimulated relaxation of a deformed polymeric glass. (PMMA, uniaxial compression up to 21% at 20°C at a strain rate of $1.7 \times 10^{-4}\ \mathrm{s}^{-1}$).

T_α (hereinafter the glass transition temperature T_g) of the polymer, but a complete recovery requires heating to T_g.

In other words, the relaxation of plastically deformed polymer glasses includes a low-temperature component at temperatures much lower than T_g (ε_{LT}) and a high-temperature component (ε_{HT}) developing in the glass transition range of the polymer (Fig. 5.4), and the total deformation of sample ε is the sum of these components

$$\varepsilon = \varepsilon_{LT} + \varepsilon_{HT}.$$

The ratio between these relaxation components is determined by the value of the preliminary deformation of the polymer ε. Figure 5.5 shows the $\sigma - \varepsilon$ diagram of the deformation of the sample (curve *1*), as well as the dependences of ε_{LT} and ε_{HT} on the strain value. It is clearly seen that for a given deformation $\varepsilon \leq \varepsilon_y$ the deformed polymer completely recovers its dimensions at temperatures below T_g due to the low-temperature component ε_{LT}. With an increase in deformation above $\varepsilon_{f.e.}$ a high-temperature component of relaxation ε_{HT} appears at a constant value of the low-temperature component.

The mechanism of the high-temperature component, as noted above, is associated with the return of deformed macromolecular coils to the initial state, which is characterized by a maximum of entropy. The nature of low-temperature relaxation is still a matter of debate. An analysis of existing views on this issue is given in the review [135]. Here we will consider only some phenomenological aspects of the problem.

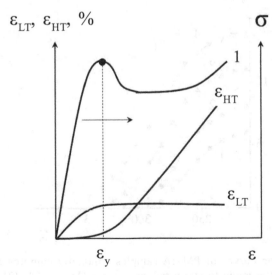

Fig. 5.5. The 'stress–strain' curve of polymeric glass (1) and the dependence of ε_{LT} and ε_{HT} on strain.

As was noted in Section 4.1, from the explanation viewpoint, the isotropic polymeric material can be used only for deformations that do not exceed the yield strain ε_{f_y} (Fig. 5.5, curve *1*). When $\varepsilon > \varepsilon_y$ plastic deformation develops according to the orientation mechanism, and the oriented polymer is localized in a macroscopic 'neck' or shear bands [107–112,145]. From operational positions, localization of deformation in heterogeneous regions means the loss of stability of the polymer properties, and the yield point is considered as the ultimate strength of the material. In this connection, we focus attention on the regularities of thermally stimulated relaxation of polymer glass deformed to $\varepsilon \le \varepsilon_y$.

Quantitatively, the nature of low-temperature relaxation is determined by

- the value of preliminary strain ε;
- deformation temperature T_{def};
- strain rate V.

For fixed T_{def} and V, an increase in ε is accompanied by a shift of the relaxation curves to higher temperatures and an increase in the temperature of the total reduction of the deformed sample, i.e. temperature at which $\varepsilon_T \rightarrow 0$ (Fig. 5.6).

For a fixed ε, an increase in the deformation temperature T_{def} leads to a similar effect (Fig. 5.7). In accordance with the principle

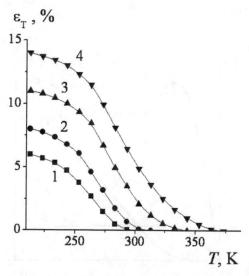

Fig. 5.6. Relaxation curves of PMMA samples uniaxially compressed at 20°C in uniaxial compression mode to strain values $\varepsilon = 6\%$ (1), $\varepsilon = 8\%$ (2), $\varepsilon = 11\%$ (3) and $\varepsilon = \varepsilon_{f.e.} = 14\%$ (4). The strain rate is 1.7×10^{-4} s^{-1}.

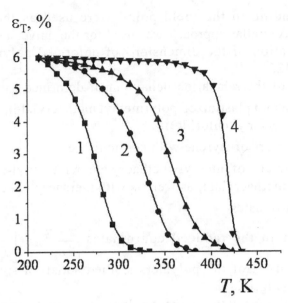

Fig. 5.7. Relaxation curves for PMMA uniaxially compressed to 6% at 20 (1), 60 (2), 80 (3), and 110°C (4).

of temperature–time superposition (see Section 1.4), the influence of the strain rate V on the shape of the relaxation curves is opposite to the effect of T_{def}.

The obtained data allow us to conclude that the evolution of the relaxation curves shown in Figs. 5.6 and 5.7 can be described in terms of the ratio of the applied strain ε and the yield strain ε_y: $\varepsilon/\varepsilon_y$

Let us explain this thesis in the following way.

For a given temperature–velocity pre-strain regime, when T_{def} and V are constant, the quantity ε_y dependent on them is also constant, and the ratio $\varepsilon/\varepsilon_y$ increases due to an increase in the given strain ε. This is manifested in the shift of the relaxation curves to the region of higher temperatures (Fig. 5.6).

For a given strain ε, an increase in the deformation temperature or a decrease in the strain rate leads to a decrease in the value of $\varepsilon_{f.e.}$, and the ratio $\varepsilon/\varepsilon_{f.e.}$ increases due to a decrease in the denominator. In this case, the result (Fig. 5.7) similar to the previous one is achieved.

In Section 4.1, it was demonstrated that the deformation behaviour of polymeric glass can be unified in the reduced, dimensionless

coordinates $\dfrac{\sigma}{\sigma_y} = f\left(\dfrac{\varepsilon}{\varepsilon_y}\right)$, where the experimentally determined

and physically justified characteristics, namely stress σ_y and strain

ε_y corresponding to the yield point, were used as the reduction parameters. A similar approach was used for the universal description of the relaxation of the dimensions of deformed polymer glasses [15,60,124,125].

Analysis of the relaxation behaviour of deformed samples
- virginh and plasticized poly(methyl methacrylate), polystyrene and poly(vinyl chloride);
- copolymers of styrene with acrylonitrile;
- copolymers of methyl methacrylate with butyl-, octyl- and lauryl methacrylate, as well as with methacrylic acid;
- polycarbonates

showed that in the reduced coordinates $\dfrac{\varepsilon - \varepsilon_T}{\varepsilon_y} = f\left[\dfrac{T - T_g}{T_g - T_{def}}\right]$ the relaxation curves of the polymers studied satisfy a single unified dependence (Fig. 5.8).

The array involved in the analysis included more than 400 independent experiments for polymers of vinyl and methacrylic series, as well as heterochain polycarbonates, previously deformed at temperatures from $-150°$ to $110°C$ in the range of strain rates from 10^{-5} to 10^{-1} s^{-1} [15,60].

Note, that the relaxation of samples of polycarbonate deformed at various temperatures and strain rates is also described by a similar (but different from that for carbon chain polymers) unified dependence.

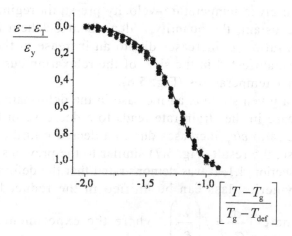

Fig. 5.8. The unified relaxation curve of polymer glasses of vinyl and methacrylic series.

Thus, for a particular class of polymeric glasses (in our case, carbon-chain and heterochain polymers), the profile of a unified relaxation curve is a characteristic function. Obviously, the laws governing the relaxation of deformed glassy polymers are determined by the relaxation nature of the preliminary deformation.

5.3. Physical ageing of deformed polymeric glasses

The physical ageing of thermodynamically non-equilibrium polymer systems (quenched, poorly annealed or deformed samples) is expressed in the instability of their properties – by changing a given parameter over time at a constant temperature. The temperature interval within which the processes of physical ageing manifest themselves most clearly coincides with the temperature interval between the temperatures of the relaxation β- and α-transitions (T_β– T_α) [20,146]. This allows us to conclude that the nature of physical ageing is associated with both quasi-independent (β-transition) and cooperative (α-transition) modes of segmental mobility (see Sections 2.1.1 and 2.1.2).

As mentioned above, deformed samples of glassy polymers are characterized by

- stored (latent) energy or excess enthalpy ΔH;
- residual strain ε_{res}.

In this regard, it is convenient to study the regularities of the physical ageing of such polymer systems by monitoring the time dependences of these parameters under the conditions of isothermal annealing.

The kinetics of enthalpy relaxation of deformed polystyrene samples was studied in detail in Refs. 20 and 147. It was shown that the annealing of polystyrene samples with ε_{res} = 5, 12, 22, 40% at 70°C (i.e. at a temperature of 10–15 degrees below T_g) leads to complete relaxation of the excess enthalpy ($\Delta H \rightarrow 0$), and in the coordinates $\Delta H = f(\ln t)$ this process is described by a dependence that is close to linear.

The data of isothermal relaxation of the residual deformation of polymeric glasses are given in [60,126]. As an example, consider the corresponding kinetic dependences for the sample of deformed poly(methyl methacrylate) (Fig. 5.9)

At temperatures below $T_g \sim 120°C$, the relaxation of deformed samples proceeds in two stages (Fig. 5.9, curves *1–5*). At the initial instant of time, the relaxation process proceeds at a high rate, after

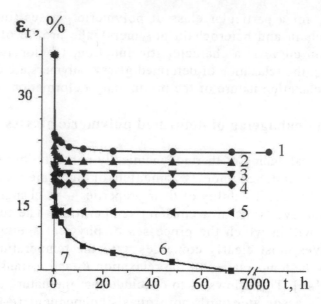

Fig. 5.9. Kinetic relaxation curves of residual strain ε_t of PMMA samples uniaxially compressed to 30% at 20°C. The annealing temperatures are 20 (1), 40 (2), 60 (3), 80 (4), 100 (5), 110 (6) and 120°C (7).

which a relatively slow yield of the values of the residual strain to a quasi-stationary value characteristic for each test temperature is observed. At temperatures close to T_g (Fig. 5.9, curves *6* and *7*), a complete restoration of the size of the deformed sample is observed. The observed behaviour is typical for a wide range of glassy polymers, regardless of their chemical structure and the magnitude of deformation.

Thus, the low-temperature (at $T < T_g$) relaxation of the residual deformation for polymer glasses is quasi-equilibrium.

From operational positions, the result obtained allows us to predict the shape stability of polymer materials and products based on them, subjected to plastic deformation during processing or physical modification.

From the point of view of the fundamental structural-deformation pattern of the behaviour of polymer glasses, the quasi-equilibrium nature of isothermal relaxation is evidence of a unique correspondence between the change in the strain $\Delta\varepsilon$ as the temperature is increased by ΔT.

This, in turn, means the existence of the distribution function $\Delta\rho(T) = \Delta\varepsilon/\Delta T$, which characterizes the temperature coefficient of relaxation of the deformed polymer at $T < T_g$ and is a stable characteristic of the material. The limit of the ratio $\Delta\varepsilon/\Delta T$ for $\Delta T \rightarrow$ 0, i.e. $\rho_\varepsilon(T) = \lim_{\Delta T \to 0} \Delta\varepsilon / \Delta T$ is a partial deformation or deformation mode that relaxes at temperature T. The function $\rho_\varepsilon(T)$ itself is, in fact, the spectrum of deformation modes distributed over the annealing temperature T.

The data of isothermal annealing of deformed polymeric glass (Fig. 5.9) make a conceptually significant refinement in modern concepts of the nature of the glassy state, according to which glass is a thermodynamically non-equilibrium but kinetically stable solid aggregate state. The results obtained make it possible to assert that the deformed polymeric glass is only partially stable in the wide temperature range below the glass transition temperature. At a fixed temperature, only a portion of the volume of the deformed glass remains stable, while the rest is relaxed.

In other words, at any temperature lower than the glass transition temperature, rigid (kinetically stable) and soft (relaxing) components coexist in the deformed glass. As the temperature rises, the ratio between these components changes, which is expressed in decreasing the residual strain.

Deformation is a macroscopic quantity related to the whole volume and, consequently, its change is possible only if there are no structural obstacles to its change in the volume. Obviously, this is possible only if the soft component is a percolated region, with respect to which the rigid part acts as an 'filler'.

When the temperature is raised by ΔT, the kinetic stability of a part of the volume ΔV is lost, which determines the relaxation of a part of the strain $\Delta\varepsilon$. This means that devitrification of the material has occurred in the volume ΔV and

$$\Delta\varepsilon(\Delta T) = \zeta(\Delta T)\Delta V(\Delta T),$$

where $\zeta(\Delta T)$ is the coefficient of proportionality, which can be defined as the deformability of the volume ΔV.

Passing to the derivatives with respect to temperature, we obtain:

$$\rho_\varepsilon(T) = \zeta(T)\rho_V(T),$$

where the spectrum of the deformation modes $\rho_\varepsilon(T)$ is related to the spectrum of volume fractions of the polymer structure $\rho_V(T)$ that devolve at the temperature T.

6

The 'brittleness–plasticity' transition and the low-temperature boundary of plasticity of polymers

The study of the laws and nature of the 'brittleness–plasticity' transition, i.e. transition from elastic to plastic deformation or from the brittle to plastic fracture mechanism is an urgent problem of polymeric material science. The plasticity of polymers determines the manifestation of the most valuable properties of this class of materials, first of all, high impact strength. In this regard, embrittlement means for a given polymer material or product on its basis a loss of the required performance characteristics and, accordingly, consumer demand.

This transition is distinguished by a pronounced relaxation character, which is not difficult to demonstrate in the following way.

Consider the mechanical behaviour of polymeric glass as a model object. Comparison of the 'stress–strain' diagrams obtained at different temperature–rate deformation modes (Fig. 6.1) indicates that the transition from plastic deformation (the family of curves $T_4 - T_6$ and $V_4 - V_6$) to brittle failure (family of curves $T_1 - T_3$ and $V_1 - V_3$) takes place at a certain superposition of temperature and strain rate.

The molecular–kinetic interpretation of these phenomena is given in Section 4.1. Here we only note the following.

The plasticity or polymeric glass develops due to segmental mobility, activated by the joint action of thermal and force fields. From relaxation positions, the condition for microscopic translational movements of segments is the inequality $t \gg \tau$, where t is the time

Brittle fracture criterion

$$\left(t \sim \frac{1}{V}\right) \ll \left(\tau \sim \frac{1}{T}\right)$$

Plasticity criterion

$$\left(t \sim \frac{1}{V}\right) \gg \left(\tau \sim \frac{1}{T}\right)$$

Brittleness–plasticity transition criterion

$$\left(t \sim \frac{1}{V}\right) \sim \left(\tau \sim \frac{1}{T}\right)$$

V – strain rate
T – deformation temperautre

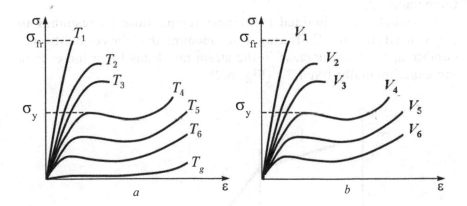

Fig. 6.1. Typical 'stress–strain' diagrams of polymeric glass at a constant strain rate (V = const) and temperatures $T_1 < T_2 < T_3 < T_4 < T_5 < T_6 < T_g$ (*a*) and at a constant deformation temperature (T = const) and strain rates $V_1 > V_2 > V_3 > V_4 > V_5 > V_6$ (b).

of mechanical action inversely proportional to the strain rate, τ is the time of transition of the segment from one state to another or the relaxation time, inversely proportional to the deformation temperature. Brittle failure is observed when $t \ll \tau$.

The sign of the two indicated inequalities corresponds to the 'brittleness – plasticity' transition can be achieved in two ways

- change in t due to a change in the strain rate with a constant deformation temperature (τ = const);
- a change in τ due to a change in the deformation temperature with constant strain rate (t = const).

The obvious condition for the 'brittleness – plasticity' transition is the comparability of the relaxation times and the external mechanical action $t \sim \tau$.

The experimental data presented in Fig. 6.1a allow us to construct the dependence of the stress of the yield stress σ_y and the brittle fracture stress σ_{fr} on the deformation temperature at a constant strain rate V = const (Fig. 6.2).

A decrease in the deformation temperature is accompanied by a linear increase in the value of σ_y (Fig. 6.2, dependence *2*) (see Section 4.1). The same behaviour is also characteristic of the brittle fracture stress σ_{fr} (Fig. 6.2, dependence *1*). In the transition from plasticity to brittleness, a clearly pronounced inflection appears on the graph, the abscissa of which is referred to as the brittleness temperature T_{br}.

Obviously, the obtained brittleness temperature corresponds to a given strain rate V. Taking into account the above-mentioned condition, $\tau \sim t$, an increase in the strain rate leads to an increase in the experimentally fixed T_{br} (Fig. 6.2).

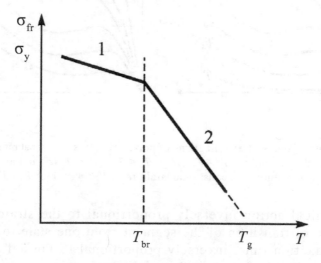

Fig. 6.2. The temperature dependences of the brittle fracture stress σ_{fr} (1) and the yield stress σ_y (2) at a constant strain rate.

As was shown in Section 4.1, the extrapolation of the linear dependence of $\sigma_{f.e} = f(T)$ to zero stress $\sigma_{f.e}$ gives the value of the glass transition temperature T_g of the polymer (Fig. 6.2), above which the deformation of the amorphous polymer proceeds by the mechanism of high elasticity.

Thus, the temperature range of the plastic deformation of polymeric glasses ranged from the brittleness temperature to the glass transition temperature of the polymer.

The above regularities and correlations are also characteristic of semi-crystalline polymers. The macroscopic picture of the influence of temperature–rate regimes on the deformation behaviour of this class of polymeric materials is analogous to that for polymer glasses (Fig. 6.1). The principal difference lies in the mechanism of deformation for semi-crystalline and amorphous, glassy polymers. Plastic deformation of a semi-crystalline polymer at temperatures much lower than its melting point T_m includes clearly fixed structural transformations of the crystalline phase. When the yield tooth is reached on the dynamometric curve, the initial crystalline (lamellar, spherulitic, etc.) structure of the material is destroyed. At $T < T_m$, this process is interpreted as 'mechanically activated melting' under the action of two fields – temperature and mechanical ($kT + \gamma\sigma$). Under deformation conditions, the 'mechanically melted' crystalline phase forms a new fibrillar crystalline modification. Formally, these structural rearrangements are described in terms of the 'recrystallization' of a semi-crystalline polymer. In connection with this, for these materials, the mechanical stress corresponding to the peak on the stress–strain diagram is called the recrystallization stress (see Section 4.2). These processes include the movement of certain kinetic units (segments and supersegmental formations) that require a certain time τ. Thus, the possibility of mechanically activated recrystallization is also determined by the ratio τ/t, which determines the relaxation character of the plastic deformation of the semi-crystalline polymer and the 'brittleness–plasticity' transition.

From the operational positions, the brittleness temperature T_{br} of the material is interpreted as a low-temperature plasticity boundary. The ambiguity of this characteristic is obvious – in each particular case this transition is realized with a certain combination of temperature and strain rate and is a kinetically dependent process. In this regard, the tabulated values of this engineering parameter are measured using strictly standardized test procedures.

As an operational characteristic, the brittleness temperature of a given polymer, determined under standard conditions, also depends on the following factors

1. When testing in shear and compression modes, the values of T_{br} are lower than those for tension;

2. For semi-crystalline polymers, T_{br} increases with increasing degree of crystallinity;

3. Uniaxial orientation extraction reduces T_{br} in the subsequent test along the axis of the preliminary orientation, and this parameter increases with the perpendicular test.

We note that the plasticization of the polymer has little effect on T_{br}, significantly reducing the glass transition temperature of the material, which is accompanied by a decrease in the temperature range $T_{br}-T_g$.

In the series of studies generalized in the monograph [20], the kinetics of plastic (inelastic) deformation of glassy polymers was related to the parameters of known relaxation transitions. The sum of the experimental data obtained made it possible to conclude that the plasticity of glassy polymers is possible only starting from the temperatures of the mechanically activated β-transition, for example, for PMMA with ≈ 220–250K. In other words, the low-temperature plasticity boundary of polymeric glasses was identified with the temperature of the relaxation β-transition T_β (see Section 2.1.1).

The structural basis of this concept is the cluster model of a glassy polymer (see Appendix 2), according to which the nature of the β-transition is determined by quasi-independent displacements of segments in intercluster regions with a reduced packing density and an excess of free volume. When the temperature is raised to the glass transition temperature T_g (or the temperature of the relaxation α-transition T_α), the segmental mobility acquires a cooperative character, capturing the entire volume of the sample, including densely packed clusters.

Correlations of T_β with a number of characteristic experimentally determined and calculated temperatures are indicative in favour of these representations.

For the disordered bodies, the existence of a characteristic temperature T^* close in magnitude to T_β was discussed in [30,148,149]. The use of Raman scattering, light scattering and small-angle X-ray scattering methods has allowed conclude that starting with $T^* \sim T_\beta$, quasi-independent liquid-like motions of chain segments appear and grow in glassy polymers as the temperature increases and the density and free volume fluctuations increase.

One should note the correspondence between T_β and the empirical characteristic temperature T_0 in the Williams–Landel–Ferry equation:

$$\log a_T = \log\left(\frac{\tau_T}{\tau_{T_g}}\right) = \frac{-c_1(T - T_g)}{c_2 + T - T_g},$$

where a_T is the shear factor, τ_T and τ_{T_g} are the relaxation times at the current temperature T and the glass transition temperature T_g, respectively, c_1 is the coefficient, $c_2 = T_g - T_0$,

and the Falcher–Vogel-Tamman equation:

$$\bar{\tau} = \tau_0 \exp\left(\frac{B}{T - T_0}\right),$$

where $\bar{\tau}$ is the average relaxation time, B is a constant,

as well as the temperature T_2 in the Gibbs–DiMarzio thermodynamic theory [33,34].

In the above empirical equations, the temperature T_0 is given the meaning of the quasithermodynamic limit, at which the total exhaustion of the fluctuation free volume is observed, and the conformational entropy tends to zero ($\Delta S_{conf} \to 0$). From the standpoint of molecular dynamics, this means a complete freezing of segmental mobility due to the termination of conformational rearrangements.

In the Gibbs–DiMarzio theory, the characteristic temperature T_2 is associated with the temperature of the second-order phase transition and is considered as the limiting hypothetical glass transition temperature, attainable only with infinitely slow cooling. The physical meaning of this limit is analogous to the temperature T_0 considered above, the zeroing of the conformational entropy.

The values of the characteristic temperatures T_β, T^*, T_0 and T_2 mentioned above are close to each other and lie in the range $(0.75 \pm 0.1)T_g$.

However, experimental data obtained in a number of papers [94, 143,150,151], indicate that the plastic deformation of polymers develops also at temperatures much lower than T_β.

All the above theories and concepts treat the transition of brittle fracture (or elastic deformation) to plasticity within the framework of the relaxation theory. For a deeper understanding of the nature of this transition, we consider the behaviour of the material under conditions that exclude the effect of obvious kinetic factors on the course of this process.

To solve this problem, we turn to the laws governing the relaxation of the dimensions of deformed samples of polymeric glasses (see Section 5.2).

We recall that the character of the thermally stimulated relaxation of a deformed polymeric glass is determined by the ratio of the deformation value ε and the deformation corresponding to the yield strain ε_y: $\varepsilon/\varepsilon_y$. In this case, the value of ε is given by the experimenter, and the value of ε_y depends on the test conditions, first of all, on the deformation temperature T_{def}.

Figure 6.3 shows the family of thermally stimulated relaxation of residual deformation of PMMA samples uniaxially compressed to a fixed degree of deformation ($\varepsilon = 6\%$) for different T_{def}. The temperature of the complete recovery of the dimensions of the deformed samples T_{rec} was evaluated as shown for a sample compressed at 293 K (Fig. 6.3, curve *3*).

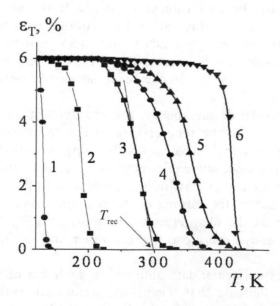

Fig. 6.3. The curves of thermally stimulated relaxation of PMMA samples uniaxially compressed by 6% at 173 (1), 253 (2), 293 (3), 333 (4), 353 (5) and 383 K (6).

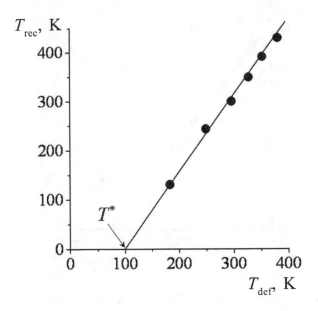

Fig. 6.4. Dependence of the recovery temperature T_{rec} on the deformation temperature T_{def} for PMMA uniaxially compressed to 6%.

The experimental results presented in Fig. 6.3, allow us to conclude that the T_{rec} value decreases linearly with decreasing deformation temperature (Fig. 6.4). Extrapolation of the obtained dependence to the zero value of T_{r3d} gives the deformation temperature T^* in the region of 100 K, the physical meaning of which is as follows.

For samples of polymethylmethacrylate uniaxially compressed to a deformation of 6% at $T > T^*$, plastic deformation develops and a residual deformation that completely relaxes at a certain temperature T_{rec} is stored. A similar procedure performed at $T < T^*$ is accompanied by elastic deformation of the polymer, and when the deformed sample is unloaded at any arbitrarily chosen temperature, a reversible restoration of its original dimensions and shape is observed.

Thus, in the case of uniaxial compression of poly(methyl methacrylate) samples by 6%, $T^* \sim 100$ K is the boundary temperature of the transition from elastic to plastic deformation.

In a similar scheme for poly(methyl methacrylate), the temperature T^* was also determined for samples uniaxially compressed by 8, 11 and 14%. The data obtained indicate that T^* decreases linearly as the

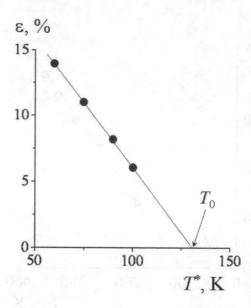

Fig. 6.5. Correlation of the temperature T^* and the value of strain ε. Poly(methyl methacrylate), uniaxial compression.

value of the given strain ε increases (Fig. 6.5), and the extrapolation of this dependence to the zero strain gives a temperature $T_0 \sim 130$ K for a given polymer.

This temperature is obtained by double extrapolation, which excludes its dependence on T_{def} and ε. From these positions, the temperature T_0 should be considered as a characteristic parameter for poly(methyl methacrylate). We emphasize that this temperature is determined without application of destructive stresses to the polymer, sensitive to the defectiveness of a particular sample, the rate of impact, etc.

Thus, in the case of uniaxial compression of poly(methyl methacrylate) samples to arbitrarily chosen degrees of deformation at arbitrarily chosen deformation temperatures, the characteristic temperature $T_0 \sim 130$ K is the boundary temperature of the transition from elastic to plastic deformation.

The question arises of extending such estimates to other polymeric glasses and searching for the relationship between the characteristic temperature T_0 and other known characteristic temperatures. To solve this problem, we turn to the unified relaxation curve of deformed polymer glasses (see Section 5.2, Fig. 5.8).

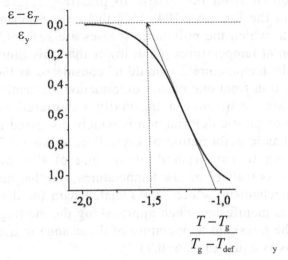

Fig. 6.6. Determination of the temperature T_0 from the unified relaxation curve of the deformed polymeric glass. Explanations in the text.

Taking into account the double extrapolation procedure used above, we process the unified curve reproduced in Fig. 6.6, as follows.

Let us draw a tangent to the steepest section of the unified relaxation curve (Fig. 6.6). In this case, the reduced temperature of complete recovery of the sample under a strain tending to zero ($\varepsilon \to 0$) is

$$\frac{T_{rec} - T_g}{T_g - T_{def}} = -1.5.$$

Taking into account that $T_0 = T_{def}$ for $T_{rec} \to 0$, we obtain

$$\frac{T_g}{T_g - T_0} = 1.5$$

or

$$T_0 = \frac{1}{3} T_g.$$

Thus, for polymeric glasses as a single class of materials, a new characteristic temperature T_0 was found and its universal relationship with the glass transition temperature of the polymer was established.

As noted above, the temperature T_0 can be interpreted as the threshold temperature of the transition from elastic to plastic deformation or from the brittle to plastic fracture mechanism. However, in the literature [94, 151] there are experimental results according to which the polymeric glasses are susceptible to plastic deformation at temperatures much lower than this limit. From these positions, the temperature T_0 should be considered as the temperature of the transition from one plastic deformation mechanism to another.

The existence of such a thermally stimulated change in the mechanism of plastic deformation is widely discussed in the general theory of plastic deformation of crystalline solids [152–156].

According to established ideas, one of the mechanisms – dislocation – is realized at low temperatures. Another mechanism, the diffusion mechanism, which is associated with the displacement of vacancies, is manifested when approaching the melting point T_m. In this case, the threshold temperature of the change in the deformation mechanisms is equal to $(0.6 \div 0.7)T_m$.

For polymeric glasses, the existence of a similar threshold temperature is discussed in [104]. The monograph [20] indicates that for amorphous polymers such a temperature is the temperature of the β-transition.

In the framework of the polycluster model of metallic glasses [157,158], the mixed mechanism of their plastic deformation, which is a superposition of the diffusion–viscous flow and the propagation of boundary dislocations into the cluster body, is noted in [159].

The studies [160,161] examined the temperature dependence of the fluctuation free volume in polymeric glasses using probes of various sizes. In the presence of an adequate fraction of the free volume, these probes are capable of conformational transitions, which are fixed in the experiment. When the temperature is lowered, the conformational mobility of the probes is 'frozen' with an ever smaller and smaller size, which quantitatively reflects a decrease in the free volume in the glassy polymer. In this case, a linear correlation was obtained between the freezing temperature of the probe and its size. Extrapolation of this linear dependence to the zero probe size gives the temperature at which the fluctuation free volume tends to zero. The values of these temperatures are for PMMA 145 K or $0.37T_g$ and for PVA 125 K or $0.38T_g$, which quantitatively corresponds to the characteristic temperature $T_0 = \frac{1}{3}T_g$ of polymeric glasses detected in the relaxation experiments.

This result suggests that the temperature T_0 is associated with the disappearance of the fluctuation free volume in the polymer. As a consequence, below this temperature, the diffusion mechanism of deformation due to the thermal and mechanical activation of translational displacements of any kinetic units can not be realized. It is reasonable to assume that under these conditions the experimentally observed plastic deformation develops according to the laws characteristic for the deformation behaviour of solids. However, clarifying the nature of this mechanism requires additional and detailed structural and mechanical studies of polymeric glasses or, in general, amorphous glassy materials at low temperatures.

7

Relaxation processes during processing of thermoplastic polymers

The processing of thermoplastic polymers (thermoplastics or plastics) is based on a complex set of technological processes aimed at the production of materials and products of a given shape, structure and the operational properties [162–167]. At present, scientific and technical principles for obtaining a wide range of required materials have been well developed by processing thermoplastics through a viscous, high elastic and vitreous state using basic techniques such as

- extrusion;
- casting;
- calendering;
- bulk and sheet stamping;
- pneumatic and vacuum forming;
- rolling, etc.

Each specific technological procedure is characterized by a specific features of the chemical, physical and physico-chemical transformations that accompany processing and determine the formation of a complex of final properties of the material. However, in all cases, processing includes

1. effect on the polymer temperature and mechanical load, or, in other words, thermal and force fields;
2. the course of relaxation processes, the completeness of which has a significant effect on the subsequent behaviour of the resulting material or article.

In this chapter, the features of both these factors are considered using the example of monolithization of powders of thermoplastic polymers under their compression moulding, that is compression in the press mould. Such systems are a convenient model for studying the whole range of the physico-mechanical processes in a polymer body, allowing in a single experiment to perform deformations of uniaxial compression and stretching, shear and hydrostatic compression.

The monolithization of the polymer powders with a particle size not exceeding 0.25 mm was carried out in a mould with a diameter of 10 mm and a height of 20 mm using two modes [143]:

'P–T' mode, which included the following steps:

1. compression in a mould of a powdered polymer at room temperature at a rate of 0.1 mm/min to a monolithization pressure of P_m;
2. heating of the compressed sample to the test temperature of monolithization T_m;
3. sample holding at given P_m and T_m for 15 minutes;
4. cooling the sample to room temperature;
5. release from the mould.

'T–P' mode, for which the step-by-step procedure included:

1. heating the powdered polymer in a mould to T_m;
2. compression in the mould of the heated sample at a speed of 0.1 mm/min to a certain monolithization pressure P_m;
3. sample holding at given P_m and T_m within 15 minutes;
4. cooling the sample to room temperature;
5. release from the mould.

The depth of monolithization was estimated from the transparency of samples liberated from the mould at room temperature. The accuracy of the 'temperature–pressure' parameter estimation, at which the transition from opacity to transparency is observed, was 5–7%.

For a deeper understanding of the essence of the relaxation phenomena occurring during the monolithization of powdered polymers, thermally stimulated recovery of the dimensions of the formed samples was studied (see Section 5.2).

To this end, in step 4 of the above modes, the moulded sample was cooled with liquid nitrogen to −60°C and at this temperature was released from the mould. The height of the sample under these conditions was taken as h_0. The resulting sample was heated at a rate of 50°/min. As a result, relaxation curves were obtained in the coordinates $\Delta h = f(T)$, where $\Delta h = h_T - h_0$ and h_T is the sample height at a given temperature.

Below are the results concerning the monolithization of polymer powders in the two used regimes [143,168–171].

7.1. Monolithization of powders in the 'P–T' mode

Visual estimates of the transparency of the samples obtained with different combinations of pressure and temperature made it possible to construct diagrams for the monolithization of powders of thermoplastics. As an example, consider a similar diagram for samples poly(methyl methacrylate) (Fig. 7.1).

This diagram is characterized by the presence of four regions (I – IV) within which samples with different physical properties are formed.

Wherein

- curve *1* separates the region I from the region II;
- curve *2* – area III from regions I and II;
- curve *3* – region IV from region III.

The samples formed in the regions I and II are opaque at room temperature, indicating that there are visible interfaces between the powder particles. Consequently, under these moulding conditions monolithization of the powder is not achieved.

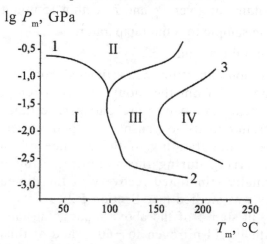

Fig. 7.1. The diagram of monolithization of powders of poly(methyl methacrylate) in the *P–T* mode. Explanations in the text.

The samples formed in region III are transparent at room temperature, i.e. monolithic, but lose this property when heated to the glass transition temperature T_g of the polymer. In other words, in the temperature range $T \geq T_g$, thermally stimulated processes occur in the sample, accompanied by a relaxation of deformation of the powder particles, a violation of close contact between them, restoration of visible interfaces and, as a result, loss of monolithicity.

The samples formed in region IV are transparent, and, therefore, are monolithic throughout the operational temperature range.

More detailed information on the behaviour of the samples obtained in the indicated regions of the *P–T* diagram (Fig. 7.1) gives the results of thermally stimulated relaxation of their sizes (Fig. 7.2).

Recall that in carrying out these studies, the moulded samples were removed from the mould at −60°C, at this temperature, their transparency was evaluated, and then heated at a constant rate.

The samples formed in region I are opaque at room temperature and at a temperature of −60°C. This means that under moulding conditions at given pressures and temperatures, the monolithization of the powder at the expense of 'splicing' of its particles and the disappearance of the interfaces between them does not occur. Compression of the powdered system is achieved only due to the elastic deformations of the particles of the original powder that instantly relax upon unloading.

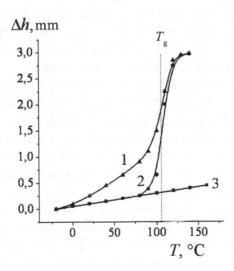

Fig. 7.2. The relaxation curves of the sizes of samples of poly(methyl methacrylate) formed in region II (1), III (2) and IV (3).

The samples formed in region II and unloaded at $T = -60°C$ are transparent. This indicates that in this region of the $P–T$ diagram the compression of a polymer powder leads to its monolithization and the disappearance of visible interfaces between the particles. However, when such samples are heated (Fig. 7.2, curve *1*), even at low temperatures ($-20÷80°C$), their dimensions are restored due to the relaxation of the deformed powder particles, which is accompanied by a loss of transparency. This relaxation process terminates at T_g of the polymer ($\sim100–110°C$) and leads, ultimately, to the separation of the samples into the original particles.

The samples formed in region III are transparent both at room temperature and at a temperature of $-60°C$. As noted above, they retain their transparency to the glass transition temperature of the polymer, at which a sharp recovery of their dimensions occurs (Fig. 7.2, curve *2*), which determines the loss of monolithicity of the product.

The samples formed in region IV are transparent under any conditions and do not show relaxation transitions in the entire studied temperature range (Fig. 7.2, curve *3*).

These data, together with independent results describing the deformation behaviour of monolithic and powdery thermoplastics during their compression moulding, have made it possible to detail the $P–T$ diagram for poly(methyl methacrylate) (Fig. 7.1) from the viewpoint of the deformation mechanism of the powder in one or another pressure region and temperatures (Fig. 7.3).

In the region I, elastic deformation of the powdery system is realized, which does not lead to an interdiffusion of the matter through the interface between the particles. Obviously, monolithization is not observed at the same time.

In region II, the glassy polymer powder undergoes yield deformation, and the system monolithizes due to the mechanically activated transfer of segments through the interfaces of the powder particles. However, such a 'pseudomonolith' remains stable only at very low temperatures, for example, at $-60°C$ in the experiments described above. As in the case of a plastically deformed block polymer (see Section 5.2), the relaxation of the plastic deformation of the powder includes a low-temperature component (at $T < T_g$) (Fig. 7.2, curve *1*), which determines the return of the deformed macromolecular coils to the initial state, restoration of the interfaces and, as a result, the loss of monolithicity. Note that in this situation the internal stresses accumulated in the polymer during the compression mouldin play an important role.

In region III, the powdered polymer is in a rubbery state. Under the influence of external pressure, segments of macromolecules easily diffuse through the interfaces of particles, thus ensuring the monolithization of the material. Chilled under load and the liberated sample is stable at $T < T_g$ for an unlimited time, but at temperatures close to the glass transition temperature, it loses monolithicity. This effect is associated with the thermal activation of segmental mobility, which ensures the relaxation of the 'frozen' high-elastic deformation (Fig. 7.2, curve *2*).

In region IV, a viscous flow state is realized for the polymer powder. Compression in this case causes a mutual diffusion of macromolecular coils in the entire volume of the sample, and monolithization proceeds by transferring the coils through the interfaces of the particles. Such a deformation of the flow is irreversible, which is responsible for the stability of the true monoliths obtained in this region of the *P–T* diagram throughout the investigated temperature interval (Fig. 7.2, curve *3*).

Thus, at a chosen monolithization pressure, for example, $P_m = 0.1$ GPa (Fig. 7.3) the increase in the temperature of monolithization T_m determines

- transition from region I to region II, i.e. transition from elastic

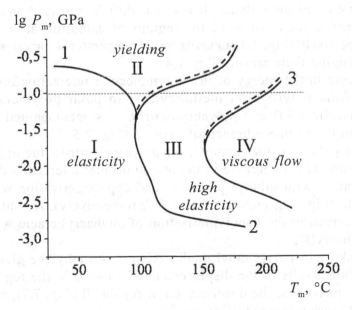

Fig. 7.3. A detailed diagram of the monolithization of powders of poly(methyl methacrylate) in the *P–T* mode. Explanations in the text.

to deformation yielding. From these positions, curve *1* is the temperature dependence of the yield stress of the powdered polymeric glass;

- transition from region II to region III, i.e. the transition from yielding to high elasticity, which is accompanied by an 'annealing' of the low-temperature component of the deformation relaxation of the powder (Fig. 7.2, the transition from curve *1* to curve *2*). From this point of view, the upper part of curve *2*, shown by the dotted line (Fig. 7.3), makes sense of the baric dependence of the glass transition temperature;

- transition from region III to region IV, i.e. the transition from high elasticity to the viscous flow, which causes 'annealing' of the high-temperature component of the deformation relaxation of the powdery polymer (Fig. 7.2, the transition from curve *2* to curve *3*). Here, the upper part of curve *3* (Fig. 7.3) is the baric dependence of the flow temperature.

The factors determining the position of the regions of monolithization and the demarcation curves (Fig. 7.3) are as follows:

1. plasticization;
2. chemical structure;
3. cross linking.

Plasticization of the polymer leads to a shift in curves *2* and *3*, which determine the position of the regions of monolithization III and IV, respectively (Fig. 7.3) towards lower temperatures, practically without changing their profile (Fig. 7.4).

An increase in the energy of the intermolecular interaction upon transition from poly(methyl methacrylate) to polar polymers – poly(methacrylic acid) and polyacrylonitrile – is accompanied by degeneration of the upper branch of curve 2 (Fig. 7.5).

For these polymers, for example, for polyacrylonitrile, the upper branch appears with a decrease in the intermolecular interaction due to plasticization with dibutyl phthalate and copolymerization with vinyl acetate (Fig. 7.6, curves *2″* and *2‴*, respectively). A similar result is observed in the copolymerization of methacrylic acid with methyl methacrylate.

Cross-linking of poly(methyl methacrylate) with ethylene glycol dimethacrylate results in the displacement of curve *2* in the region of higher temperatures, the degeneration of region III (Fig. 7.7), and the complete disappearance of region IV.

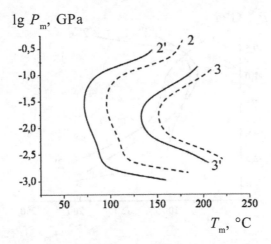

Fig. 7.4. Curves 2 and 3 (dashed lines) are poly(methyl methacrylate). Curves 2'and 3' (solid lines) are poly(methyl methacrylate), plasticized with dibutyl phthalate (9 wt.%). The designations of the curves are the same as in Fig. 7.3

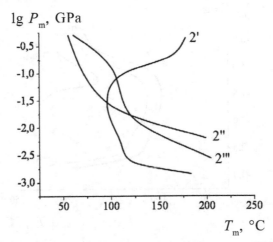

Fig. 7.5. Curves 2, determining the position of the region of monolithization III, for poly(methyl methacrylate) (2'), poly(methacrylic acid) (2″) and polyacrylonitrile (2‴),

7.2. Monolithization of powders in the T–P mode

The transition from the P–T mode (heating of the precompressed sample) to the T–P mode (compression of the preheated sample) does not qualitatively change the monolithization pattern presented in the previous section. The quantitative difference between these cases is that in the T–P mode, the upper branches of curves *2* and

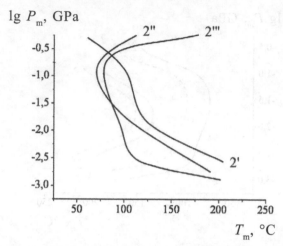

Fig. 7.6. Curves 2 for polyacrylonitrile (2′), polyacrylonitrile, plasticized with dibutyl phthalate (9 wt%) (2″) and copolymer of acrylonitrile/vinyl acetate (85/15) (2‴).

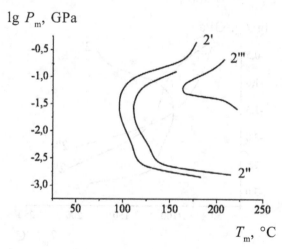

Fig. 7.7. Curves 2 for poly(methyl methacrylate) (2′) and poly(methyl methacrylate), crosslinked with dimethacrylate with ethylene glycol. Concentration of the crosslinker: 1.5 (2″) and 8 mol% (2‴).

3, bounding the regions of monolithization III and IV, are shifted toward lower temperatures (Fig. 7.8).

The observed difference is due to the fact that in the $P–T$ mode, the compression pressure is applied to the glassy polymer powder, and in the $T–P$ mode – to the polymer powder either in the high elastic or in the viscous flow state. In the latter variant, the processes responsible for the monolithization of the system due to the diffusion

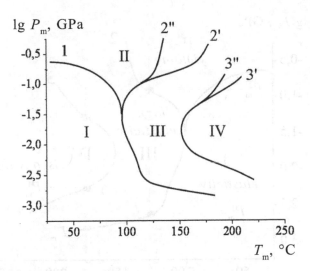

Fig. 7.8. The diagram of monolithization of powders of poly(methyl methacrylate). *2'* and *3'* – the *P–T* mode. *2"* and *3"* – the *T–P* mode.

of segments and macromolecular coils through the interface between the particles occur just in the course of loading.

The relaxation nature of the phenomena associated with the migration of kinetic units can be easily demonstrated as follows. With an increase in the compression rate, the upper parts of the curves *2"* and *3"* (the *T–P* mode) tend to the curves *2'* and *3'* (*P–T* mode) (Fig. 7.8). Obviously, the coincidence of the curves obtained in these monolithic regimes should be expected when the compression rate (time) becomes comparable with the relaxation rate (time).

It is remarkable that the upper branches of curves *2* and *3* exist when the polymer is formed in the *T–P* regime. It would seem that an increase in the pressure applied to the powdered polymer at temperatures above the glass transition temperature should in any case contribute to the monolithization of the system. This argument, however, is contradicted by the experimental results shown in Fig. 7.8. Let us clarify the situation by examining in detail the diagram of monolithization in the *T–P* mode (Fig. 7.9).

At the monolithization temperature $T_m^1 > T_g$(~100–110°C), the original powdered polymer is in a rubbery state.

Compression of a rubber-like powder with a pressure $P_m < P_m^1$ (region I) is accompanied by elastic and reversible deformation of the particles and does not lead to the formation of monolithic samples.

In the pressure range $P_m^1 < P_m < P_m^2$ (region III), the thermally activated segmental mobility ensures the monolithization of the

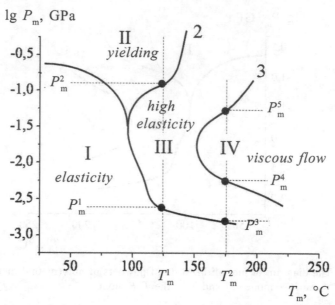

Fig. 7.9. The diagram of monolithization of powders of poly(methyl methacrylate). T–P mode. Explanation in the text.

system due to the diffusion of the segments through the interfaces between the powder particles by the mechanism of high-elastic deformation. In this region, samples that retain monolithicity to T_g of the polymer are formed (Fig. 7.2, curve *2*).

Compression of the powder with pressures exceeding the value of P_m^2 ($P_m > P_m^2$)(area II) leads to a significant compaction of the system, a decrease in the fraction of the free volume and, as a result, to the glass transition of the polymer. The deformation of such mechanically vitrified powders is carried out by the mechanism of yielding, and their monolithization proceeds by mechanically activated segment transfer through the interfaces of the particles. The obtained samples lose their monolithicity even at temperatures much lower than T_g of the polymer due to the low-temperature relaxation of the plastic deformation (Fig. 7.2, curve *1*).

At a monolithization temperature of $T_m^2 > T_m^1$ compression in the pressure ranges $P_m < P_m^3$ (region I) and $P_m^3 < P_m < P_m^4$ (region III) is described by the same laws as the previous case.

The pressures lying within the $P_m^4 < P_m < P_m^5$ range, (region IV) cause irreversible deformation of the flow and ensure the monolithization of the powder due to the diffusion of macromolecular coils through the interfaces between the particles. This causes the formation of true monoliths that are stable at any operating temperatures (Fig. 7.2, curve *3*).

At pressures $P_m > P_m^5$, the free volume fraction decreases, which prevents large-scale displacements of macromolecular coils. This determines the transition to the high-elastic deformations characteristic of region III.

Appendix 1

Thermomechanical analysis

Thermomechanical analysis (TMA) is a set of methods for studying the physical and mechanical behaviour of materials under the combined action of force and thermal fields. These methods are divided into two groups, differing in the nature of the application of mechanical stress to the studied sample – static and dynamic regimes. Below are considered the basic principles of these options.

1. Static thermomechanical analysis

This experimental approach is based on the following procedure.

At temperature T_1, the sample is subjected to a constant mechanical stress σ for a certain time t and the strain ε developed under the given conditions is fixed. After this, the stress is removed and the experiment is repeated at T_2 temperature and at the same stress and exposure time. The conditions for thermostatic control, the magnitude and time of the force action, as well as the temperature difference between the two adjacent measurements ($\Delta T = T_2 - T_1$), remain constant throughout the experiment. This requires that the current mechanical load does not cause structural changes in the sample and does not lead to the accumulation of irreversible strains.

In another version of the static thermomechanical analysis, a permanent mechanical stress is used.

The result of the experiment is a thermomechanical curve – dependence of the strain on the temperature $\varepsilon = f(T)$, a typical form of which for an amorphous polymer is shown in Fig. A1.1.

The pronounced inflections associated with a sharp increase in deformation in a narrow temperature range are interpreted as transitions from the vitreous to the high elastic state at the glass

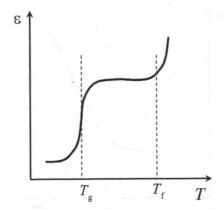

Fig. A1.1. A typical thermomechanical curve of an amorphous polymer.

transition temperature T_g and from the rubbery to the viscous flow state at a flow temperature T_f. More generally, the thermomechanical research method allows one to track the progress of any chemical and physical processes that are accompanied by a change in the deformability of the sample (crystallization, melting, crosslinking, curing, softening, etc.).

Dilatometry is considered as a variant of static mechanical analysis – a method based on measuring the linear dimension of a sample l or its volume V with a temperature change without an external mechanical load ($\sigma = 0$).

As a result, determine the linear $\left(\alpha = \dfrac{dl}{l}\dfrac{1}{dT}\right)$ and the volume $\left(\beta = \dfrac{dV}{V}\dfrac{1}{dT}\right)$ thermal expansion coefficients. Changes in these coefficients as a result of chemical reactions in the material, structural transformations, phase and physical transitions allow us to estimate the temperature intervals of these processes. Thus, for example, the dilatometric curve of an amorphous polymer in the coordinates $V = f(T)$ (Fig. A1.2) is characterized by a pronounced fracture at the glass transition temperature, which is associated with a sharp (by a factor of 3–4) increase in β.

2. Dynamic thermomechanical analysis

The basis of this version of thermomechanical analysis is the cyclic effect on the sample in the following modes:

1. Load up to a given stress value σ and unloading to $\sigma = 0$;
2. Extension to a given value of strain ε and reduction to $\varepsilon = 0$.

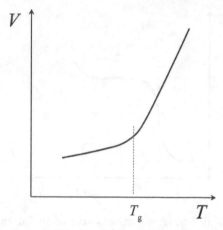

Fig. A1.2. A typical dilatometric curve of an amorphous polymer.

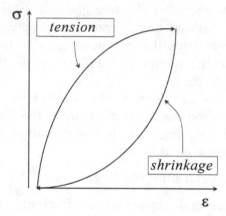

Fig. A1.3. A typical diagram of a cyclic test in the «stretch – shrink» mode.

An indispensable condition for the test is a complete return of the sample to its original state, i.e. zeroing both stress and strain. The result of both one and the other mode is the 'stress–strain' diagram, the typical form of which is shown in Fig. A1.3.

The processing of the obtained diagrams is reduced to the evaluation of the following parameters:

- the area under the extension curve S, which is the deformation work per unit volume of the sample.

$$S = \int_0^\varepsilon \sigma d\varepsilon, \text{ where } \sigma d\varepsilon = \frac{f}{S_{c.s.}} \frac{dl}{l_0} = \frac{fdl}{V} = \frac{A_d}{V},$$

where f is the force, $S_{c.s.}$ is the cross-sectional area of the sample, V is the initial volume of the sample, and A_d is the deformation work.

- the area under the contraction curve, which characterizes the part of the spent work A_{el}, which the physical body reversibly (elastically) returns when unloading.
- the area of the hysteresis loop, which determines the part of the work expended, which is irreversibly 'lost' during the cyclic test, A_{loop}.

- the coefficient of mechanical losses $\chi = \dfrac{A_{loop}}{A_d}$.

The nature of hysteresis phenomena or mechanical losses observed for viscoelastic bodies in the cyclic action regime is associated with translational movements of one or another type of kinetic units (molecules in low molecular weight bodies, segments in polymers). As a result, internal friction occurs in the sample, and mechanical work is partially dissipated as heat. In contrast, the elastic deformation of the Hookean body is determined by the deviation of the kinetic units from the equilibrium position without translational displacements. At the same time, internal friction does not occur there, there are no mechanical losses, and there is no hysteresis ($\chi = 0$). When the ideal fluid is deformed (viscously flowing), the entire work expended for deformation is completely 'lost' ($\chi = 1$).

In instrumental practice, dynamic mechanical analysis involves applying sinusoidally varying strain or stress to a sample.

If a sinusoidally changing strain

$$\varepsilon = \varepsilon_0 \sin \omega t$$

is applied to an ideally elastic body, then according to Hooke's law the resultant stress is expressed as

$$\sigma = E\varepsilon = E\varepsilon_0 \sin \omega t = \sigma_0 \sin \omega t.$$

This means that there is no phase lag between the applied strain and the resultant stress. In other words, the elastic body instantly reacts to the external action.

If the sinusoidally varying stress

$$\sigma = \sigma_0 \sin \omega t$$

is applied to an ideal fluid, then according to Newton's law the resultant deformation is expressed as

$$\sigma = \eta \frac{d\varepsilon}{dt} \Rightarrow \frac{d\varepsilon}{dt} = \frac{\sigma_0 \sin \omega t}{\eta}.$$

After integration we obtain

$$\varepsilon = \varepsilon_0 \sin\left(\omega t - \frac{\pi}{2}\right),$$

where $\varepsilon_0 = \sigma_0/(\eta\omega)$.

Consequently, for the Newtonian fluid, the resulting strain lags behind the applied stress by a phase angle equal to $\pi/2$.

The physico-mechanical behaviour of an amorphous viscoelastic body is characterized by a combination of elastic and viscous responses. It is reasonable to assume that in the case of application of a sinusoidally varying strain

$$\varepsilon = \varepsilon_0 \sin \omega t$$

the resultant stress will lag behind the strain for $0 < \delta < /2$:

$$\sigma = \sigma_0 \sin(\omega t - \delta).$$

Schematically, this situation is shown in Fig. A1.4.
In the Cartesian coordinates, the sinusoidally changing strain ε can be represented as the rotation of the deformation vector around zero with the angular frequency ω (Fig. A1.5). In this case, the phase lag

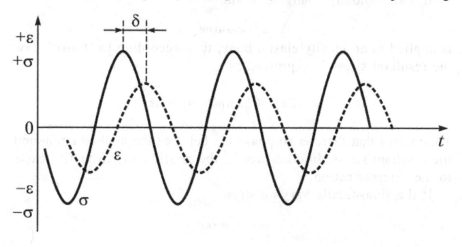

Fig. A1.4. Sinusoidal stresses σ and strains ε for an amorphous viscoelastic body during cyclic tests.

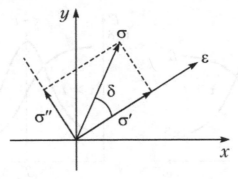

Fig. A1.5. The sinusoidal variation of the applied strain ε and the resultant stress σ in Cartesian coordinates.

between the applied strain and the resultant stress σ is expressed as the angle δ between the strain and stress vectors.

We divide the resultant stress vector into two components: σ', which is in phase with deformation, and σ'', which is out of phase. Then, the resultant stress can be represented as the sum of the real and imaginary components: $\sigma^* = \sigma' + \sigma'' i$. Normalizing the stress by the amount of deformation, we obtain

$$\frac{\sigma^*}{\varepsilon} = \frac{\sigma'}{\varepsilon} + \frac{\sigma''}{\varepsilon} i \Rightarrow E^* = E' + E'' i,$$

where E^* is a complex modulus; E' is the elastic or storage modulus; E'' is the loss modulus.

The storage modulus E' characterizes a part of the mechanical work that accumulates in the sample in the form of elastic energy and reversibly returns when unloading. The loss module E'' describes the part of the mechanical work that is irretrievably dissipated in each cycle as heat, and their ratio is called the loss tangent tg $\delta = \sigma''/\sigma' = E''/E'$.

Figure A1.6 shows for a viscoelastic body typical dependences of E' and tg δ on the frequency of the action at a constant test temperature (Fig. 1.6a) and their temperature dependences at a constant exposure frequency (Fig. A1.6b).

Within the framework of the molecular–kinetic theory, viscoelasticity is determined by mutual displacements of kinetic units – molecules in the case of low-molecular-weight amorphous bodies and segments for amorphous polymers.

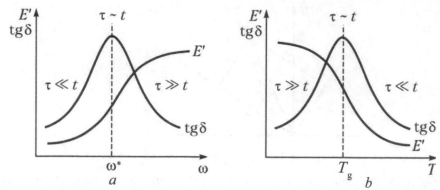

Fig. A1.6. The dependence of the storage modulus E' and the loss tangent tg δ on the frequency of the action (a) and temperature (b) for an amorphous viscoelastic body.

The constancy of the test temperature (Fig. A1.6a) means the constancy of the temperature-dependent average relaxation time of the kinetic unit τ, i.e. the time required to transfer the kinetic unit from one position to another. Increasing the frequency of exposure ω means decreasing the exposure time t.

At low frequencies $\tau \ll t$. Under these conditions, the kinetic units manage to progressively move relative to each other, which causes the deformation of the flow to develop in low-molecular-weight bodies, and in the polymers, high elasticity. In this state, the elastic modulus of the amorphous body has low values, the deformation process is close to equilibrium, and as a result, the mechanical losses are small.

At high frequencies $\tau \gg t$. Accordingly, the kinetic units do not have time to move from one position to another, and the deformation is due to their small displacements near the equilibrium position, similar to the elastic strains of Hookean bodies. As a result, the material is characterized by a high storage modulus and low mechanical losses. The latter is due to the fact that there is no mutual displacement of the segments, internal friction does not arise and the deformation process has an equilibrium character.

In the neighborhood of the frequency ω^* the condition $\tau \sim t$ is satisfied. In this region, the process of deformation is unbalanced, which causes high values of mechanical losses. In this case, a small increase in frequency leads to a sharp change in the storage modulus.

The constancy of frequency in the cyclic test (Fig. A1.6b) means the constancy of time t. As the temperature increases, τ decreases, and, as in the previous case, the ratio of these two time parameters changes. This factor is responsible for the observed behaviour of the material, similar to the case considered in Fig. A1.6a.

The temperature at which the tangent of the loss tangent passes through a maximum ($\tau \sim t$) is the glass transition temperature. In other words, at a ratio $\tau/t \sim 1$ for an amorphous low-molecular-weight body, a transition from the vitreous to the viscous flow state of the 'thawing' of molecular mobility is observed, and for the amorphous polymer – from the glassy to the highly elastic state due to the 'thawing' of the mobility of the segments.

The relaxation nature of these transitions is easily demonstrated as follows.

Figure A1.7 shows the temperature dependences of E' and tg δ obtained at different angular frequencies. An increase in the frequency is accompanied by a shift of these dependences to higher temperatures and an increase in the experimentally determined glass transition temperature T_g. An increase in frequency means a decrease in exposure time t. In this connection, the vitrification condition $\tau \sim t$ is realized at lower relaxation times, i.e. at higher temperatures.

Thus, the results of dynamic thermomechanical analysis allow us to conclude that the mechanical response of a material is determined by the ratio of time t and the relaxation time of the kinetic unit responsible for the development of the elementary deformation act τ.

This approach is formalized in terms of the Deborah number $De = \tau/t$ and underlies the principle of temperature–time superposition (see Section 1.4), according to which the same mechanical response of the material can be achieved either by changing the exposure time

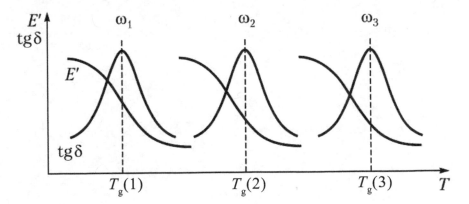

Fig. A.1.7. The temperature dependences of the storage modulus E' and the loss tangent tg δ at frequencies $\omega_1 < \omega_2 < \omega_3$.

at a constant temperature, i.e. at a constant relaxation time, or by a change in the relaxation time with a change in temperature at a constant exposure time.

To summarize, we note that thermomechanical analysis allows

- determine the temperatures of phase, relaxation and physical transitions, as well as study their dependence on the molecular weight of the material, the history of the sample, chemical, physico-chemical and physical modification, etc.;
- to study the regularities of phase, polymorphic, structural and chemical processes occurring in the material under certain parameters of the temperature–force action;
- predict the physico-mechanical behavior of the material in the specified temperature–time modes of operation,

In this regard, this methodology is widely used both in research practice, and in industry and marketing, where standardized procedures of thermomechanical analysis serve for characterization, certification, product identification, quality control of products and their operational stability, etc.

Structure of amorphous polymers

The question of the physical structure of amorphous polymers, as, well as the structure of low-molecular-weight amorphous substances and materials, is still a matter of debate. This is due to the fact that for the study of the structure of disordered systems, diffraction methods that have proved themselves in the study of crystalline bodies are inapplicable. In many cases, conclusions about the structural organization of amorphous polymers are made on the basis of an analysis of their physico-mechanical behaviour.

The established provisions concerning the views on the physical structure of this class of polymers are reduced to the following.

First, for the amorphous polymers, the short-range order in the arrangement of fragments of macromolecules is characteristic. The size of the short-range order regions is comparable to the size of segments of flexible-chain polymers and lies within a dozen nanometers.

Secondly, the physical structure of amorphous polymers is interpreted from the standpoint of a fluctuation, physical network (Fig. A2.1).

The nodes of the physical network are the entanglements of macromolecules (Fig. A2.1b), as well as ordered clusters (λ-structures), which are constructed from segments of macrochains in folded (Fig. A2.1c), straightened (Fig. A2.1d) or globular conformation (Fig. A2.1e). Lambda structures form a fluctuation network — as a result of thermal motion, they are continuously formed and dissipated.

Consider in more detail the structure of the physical network, the nodes of which are clusters.

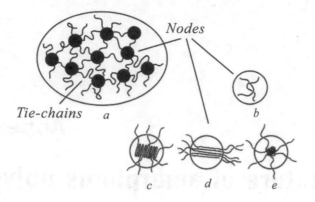

Fig. A2.1. Schematic representation of the structure of an amorphous polymer.

Each macromolecule participates in the formation of several clusters or, in other words, passes through several clusters. The sections of macromolecules that connect the clusters to each other are called tie-chains (Fig. A2.1*a*).

The cluster model of the structure of an amorphous polymer has been widely discussed in the literature during the last decades (6, 20, 157, 172–179). Leaving aside discussions on the fine structure of clusters and details of their structure formation, we shall single out the essential provisions of this model, which are necessary for further analysis of the relaxation phenomena in amorphous polymers.

So, the cluster is a supersegmental structural formation of a fluctuation origin of 5–10 nm in size, characterized by an increased local order and density in comparison with intercluster, more friable and disordered regions. The latter form the matrix of an amorphous polymer, in which the entire fluctuation free volume is concentrated.

Fluctuation decay and cluster formation involve the transition of a segment of the macromolecule from the intracluster ordered state to the intercluster space and vice versa. These elementary acts of transition from one state to another require a certain time τ, which is considered as the 'sedentary' life of the segment or the relaxation time.

According to the estimates given in [18,21], for amorphous polymers in the rubbery state, the 'settled' life of the segment within the tie-chain is $10^{-6} \div 10^{-4}$ s, and within the cluster – $10 \div 10^{4}$ s. For low-molecular-weight liquids, the time of 'sedentary' life of molecules is $10^{-10} \div 10^{-8}$ s.

Obviously, the clusters are characterized by a distribution in size, density, concentration of segments, and the like. This determines the wide spectrum of 'settled' segment life in a cluster or, in other words, the spectrum of relaxation times.

Thus, the physical structure of an amorphous polymer can be represented as a spatial molecular network with nanoscale physical nodes, whose role is played by clusters characterized by different lifetimes. Such a fluctuation network is kinetically, but not thermodynamically stable, the node lifetime decreases with increasing temperature. When an external, for example, mechanical field is applied, the behaviour of such a system depends on the ratio of the time of action t and the lifetime of the node or, what is the same, the relaxation of the segment τ.

Recently, the cluster model finds a regular development when considering polymers as natural nanostructured objects [180–187].

The above representations interpret the structure of an amorphous polymer within the framework of a two-level model that takes into account the coexistence of rigid (clusters) and soft (intercluster regions) components. An analysis of the results on the thermally stimulated relaxation of deformed polymeric glasses made it possible to conclude that there is a continuous spectrum of microheterogeneous structures in an amorphous polymer distributed at their local glass transition temperatures [60]. The genesis of such a multilevel structure is associated with the physical and physico-chemical features of polymerization.

In the simplest variant of polymer synthesis, namely in the case of homophase, bulk radical polymerization, the maximum achievable conversion depth at a given temperature is determined by the viscosity of the system and the diffusion rate of the particles participating in the polymerization process.

The increase in the viscosity of the polymerization system, due to the growth of polymer chains, is accompanied by an increase in its glass transition temperature. When the glass transition temperature of the system reaches the polymerization temperature, complete immobilization of the monomer molecules and growing macromolecules is observed, and the process ceases. The observed behaviour is interpreted as isothermal vitrification caused by a chemical polymerization reaction. To continue the polymerization, a rise in temperature is required.

A study of the fine mechanism of such physico-chemical transformations, using the example of poly(methyl methacrylate), has shown [188] that during the radical bulk polymerization an inhomogeneous distribution of the components (macroradicals, monomer and initiator) occurs in the polymerization system. In the region of autoacceleration, after reaching the gel point, immobilization of macroradicals in a viscous polymerization system occurs. As a result, the macroradicals become physically isolated from each other, and the chain growth reactions are localized in the neighbourhood of these immobilized centres. In other words, the polymerization system is distinguished by the presence of spatial inhomogeneities (microregions), which due to continuous initiation are formed during the entire polymerization time.

Each of these microheterogeneous regions is characterized by its degree of reaction completion, size and molecular weight characteristics, which in turn determines the uneven distribution of density and free volume in the polymerization system. As we approach 100% conversion, the difference between the microregions is smoothed out, but the uneven distribution of these characteristics is also retained in the final polymer.

Thus, the treatment of radical polymerization from the position of isothermal vitrification leads to the conclusion that a set or a spectrum of heterogeneous microregions, different in density and fraction of free volume forms in the polymer.

Structure of semi-crystalline polymers

The results of long-term studies of the structure of crystalline polymers are summarized in numerous monographs and reviews, for example, [4,46,73,81–85,133,134]. The structural and morphological types observed in polymers and materials based on them are very diverse and often specific for a given polymeric substance. Without claiming a comprehensive review of the problem, we present in this section the most general ideas about the crystal structure of polymers necessary for analyzing their relaxation properties.

In the general case, the crystalline phase state of a solid is characterized by a three-dimensional long-range order given by the crystal lattice. The repeating element of the crystal lattice is called the elementary cell, which is described by three non-coplanar vectors $\bar{a}$, $\bar{b}$, $\bar{c}$ and three angles between them α, β, γ.

Depending on the ratio of these vectors and angles, there are seven types of crystal lattice:

1. cubic,
2. tetragonal,
3. hexagonal,
4. rhombic (orthorhombic)
5. rhombohedral (trigonal),
6. monoclinic,
7. triclinic.

Polymers crystallize with the formation of the last six types. The cubic crystal lattice for polymer crystals is not known.

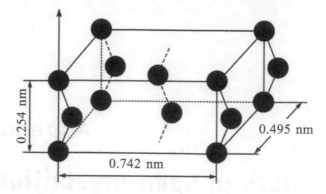

Fig. A3.1. Schematic representation of a unit cell of polyethylene: black circles – polymer –CH$_2$–groups; ↑ – direction of the axis of the macromolecule.

A specific feature of polymer crystals is that the length of the macromolecule far exceeds the dimensions of the unit cell. In this connection, the individual macromolecule takes part in the formation of a large number of elementary cells. In Fig. A3.1 this situation is illustrated by the example of the crystalline structure of polyethylene.

The polyethylene macromolecules in the low-energy trans-conformation crystallize to form a rhombic unit cell with a size of $0.742 \times 0.495 \times 0.254$ nm^3, the points of which form the –CH$_2$–groups of the polymer. The axis of the macromolecule coincides with the vector $\bar{c}$, and the hydrogen atoms lie in planes parallel to the plane *ab*. Such an elementary cell is constructed with the participation of five macromolecules: four of them form cell points, and the fifth one is in the centre. These chains form a consecutive set of elementary cells, thus forming the crystalline structure of the polymer.

We note that for most crystalline polymers, the molecular packing coefficients (the ratio of the intrinsic volume of the units of the macromolecule entering the unit cell to the total volume of the unit cell) lie in the interval $0.62 \div 0.67$ and are close to the packing coefficients of crystals of low-molecular-weight organic substances.

As for low-molecular-weight crystals, **polymorphism** is characteristic for crystalline polymers – the same substance can crystallize with the formation of various types of crystal lattice depending on the conditions of crystallization and external action. For example, at temperatures below 19°C, polytetrafluoroethylene crystallizes in a trigonal modification, and at higher temperatures it crystallizes in a rhombic one. The deformation of polyethylene with

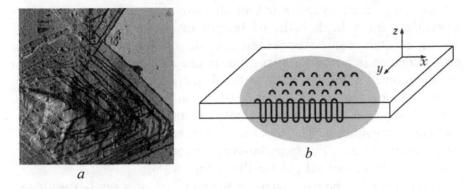

Fig. A3.2. A micrograph of a single-crystal polyethylene (*a*) and a schematic representation of the structure of the lamellae (*b*)..

orthorhombic structure is accompanied by its transformation into a monoclinic one.

For a crystalline polymer with a fixed type of crystal lattice, it is possible to form a large number of different supramolecular structures, some examples of which are given below.

The formation of single crystals is characteristic only of a limited range of polymers, for example, for the polyethylene crystallized from a dilute solution. A micrograph of a single-crystal polyethylene is shown in Fig. A3.2*a*.

Such lamellar crystals with dimensions from one to tens of micrometers (along the *x* and *y* axes) and a thickness on the *z* axis not more than two tens of nanometers are called lamellas. The length of an individual macromolecule exceeds the lamella thickness by for 2–3 orders of magnitude. The packaging of macromolecules in the lamellas is carried out by folding their chains. Schematically, the structure of the lamellas is shown in Fig. A3.2*b* (highlighted area). Along the *z*-axis, the axis of the macromolecule is directed in a folded conformation, and on the *xy*-plane all folds forming the lamella are located. The thickness of the lamellae and, accordingly, the length of the fold, depends on the choice of solvent and the crystallization temperature. For example, for polyethylene, an increase in the crystallization temperature from 50 to 90°C leads to an increase in the thickness of the lamellas from 9 to 15 nm.

An increase in the concentration of the solution and the speed of crystallization is accompanied by the formation of more complex terraced-like supramolecular structures constructed by laminating many lamellas.

As the rate of evaporation of the solvent increases, fibrillar crystals with a high ratio of length to thickness are formed. These crystals can be considered as degenerate lamellas, when crystallization is not in the plane, but in one direction. Crystallization in supercooled solutions leads to the formation of dendritic crystals.

A common feature of the crystallization of polymers is the fact that a single macrochain participates in the formation of several individual crystallites. The tie-chains connecting the crystallites form amorphous regions. For lamellar-type crystallites, a variant of such a structure is shown schematically in Fig. A3.3.

Quantitatively, the parameters of the structure of a semi-crystalline polymer are described by the crystallite size or for the case shown in Fig. A3.3, lamella thickness (fold length) L, as well as the value of a repeat period L_p, which includes lamellas and an adjacent amorphous interlayer.

When crystallization is carried out from concentrated solutions and melts the formation of spherulites is observed (Fig. A3.4a).

The size of spherulites reaches thousands of microns, and they are clearly visible when using optical microscopy. Spherulites are constructed from a set of lamellar crystallites growing from a single centre of crystallization and forming a radius of spherulite (Fig. A3.4b). The lamellas constituting the radius of the spherulite are separated by amorphous regions with the participation of tie-chains. In the radial spherulites, the radii are formed by planar lamellas. In a number of cases, during crystallization, the lamellas are twisted with the formation of spiral structures. The spherulites with spiral radii are called ringed spherulites.

Thus, the structure formation of a semicrystalline polymer is characterized by a wide range of 'degrees of freedom'. Depending on the chemical nature of the substance and the temperature–time

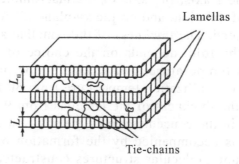

Lamellas

Tie-chains

Fig. A3.3. Schematic representation of the structure of a semi-crystalline polymer.

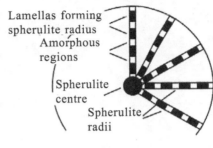

Lamellas forming
spherulite radius
Amorphous
regions

Spherulite
centre

Spherulite
radii

a

b

Fig. A3.4. A photomicrograph of polypropylene spherulite (a) and a schematic representation of the spherulite structure (b).

regimes of crystallization, the following structural parameters of this class of materials are possible:

1. the degree of crystallinity;
2. the size of the crystallites and the magnitude of the repeat period;
3. type of crystal lattice; polymorphic modifications;
4. morphology (lamellas, spherulites, fibrils, etc.).

Such a variety of quantitative and qualitative structural characteristics determines the complexity of the relaxation behaviour of semicrystalline polymers and, often, the ambiguity in the interpretation of the experimental data (see Section 2.3)

Unification of the deformation behaviour of polymer glasses

In Section 4.1, the relaxation nature of the plastic deformation of polymer glasses was found. As a result, the profile of the 'stress–strain' diagram, as well as the mechanical parameters of the material (primarily the elastic modulus E_0, stress and strain corresponding to the yield point, σ_y and ε_y, respectively (Fig. 4.1)) are controlled by the temperature–rate modes of mechanical action (Figs. 4.2 and 4.3).

The nature of these effects is due to the fact that the mechanical response of an individual polymer is controlled by the ratio of two time parameters

- τ is the relaxation time of the kinetic unit responsible for the elementary deformation act. This quantity is inversely proportional to temperature: $\tau \sim 1/T$;
- t is the time of mechanical action inversely proportional to the strain rate V and deformation frequency ω: $t \sim 1/V \sim 1/\omega$.

At a given temperature, polymers of different chemical nature are characterized by different values of τ.

Thus, within the framework of relaxation views, the deformation behaviour of a polymer glass is determined

- chemical structure;
- deformation temperature and strain rate.

As was shown in Section 4.1 for the initial section of the deformation curve limited by the yield strain, the unification of the deformation behaviour is achieved by normalizing the current values of stress σ and strain ε by the stress and strain values corresponding to the limit of forced elasticity, $\sigma_{f.e}$ and $\varepsilon_{f.e.}$, respectively, (Fig. A4.1).

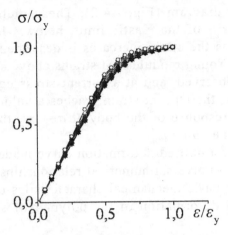

Fig. A.4.1. Unified 'stress–strain' diagram curve of polymeric glass.

Thus, in the reduced dimensionless coordinates, $\dfrac{\sigma}{\sigma_y} = f\left(\dfrac{\varepsilon}{\varepsilon_y}\right)$ the strain range $\varepsilon \leq \varepsilon_y$, i.e. the interval of practical use of the isotropic material is described by a single, universal curve, regardless of the chemical structure of the polymer, as well as temperature and time modes of deformation.

The construction of the unified deformation curve (Fig. A4.1) involved the processing of more than 500 independent experiments on the deformation behaviour of carbon-chain, heterochain and heterocyclic polymer glasses [15,60]:

- methacrylates and plasticized methacrylates;
- copolymers of methacrylic monomers of various compositions;
- poly(vinyl chloride) (PVC), plasticized PVC and mixtures based on PVC;
- polystyrene;
- polycarbonate;
- polyimides;
- epoxy resins;
- cellulose acetate,

deformed in the modes of uniaxial tension and compression, shear and torsion in the temperature range from −100 to 200°C and strain rates from 10^{-5} to 10^{-2} s^{-1}.

For a more detailed analysis of the unified deformation behaviour of polymer glasses, let us turn to the characteristics of the studied region of the σ–ε diagram (Fig. A4.2). These include the stress σ_{pr} and the strain $\varepsilon_{e.l.}$ of the elastic limit, below which the linear increase in stress as the strain increases is described by the initial modulus E_0 or the Young modulus. At strains above $\varepsilon_{e.l.}$, a deviation from linearity is observed, and at a current stress equal to $\sigma_{f.e}$, for the glassy polymer, the current strain reaches a value of $\varepsilon_{f.e}$. In the case of the linear response of the body at $\sigma = \sigma_{f.e.}$ the deformation is expected to be the value ε_{lin}.

The existence of a unified deformation curve made it possible to reveal the following universal numerical relationships reflecting the interrelation of the basic mechanical characteristics of the material and describing the deformation of the polymer glass in a unified way [15,60].

Thus, the relationship between the above parameters correlations

$$\frac{\varepsilon_{el}}{\varepsilon_y} = 0.5 \pm 0.05 \tag{A4.1}$$

and

$$\frac{\sigma_{el}}{\sigma_y} = 0.8 \pm 0.05 \tag{A4.2}$$

The ratio

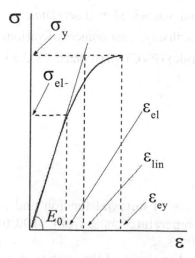

Fig. A4.2. The initial section of the 'stress–strain' curve of the polymeric glass. Explanations in the text.

$$E^* = \frac{\sigma_{el}}{\sigma_y}\frac{\varepsilon_y}{\varepsilon_{el}} = E_0\frac{\varepsilon_y}{\sigma_y} = 1.6 \pm 0.2 \qquad (\text{A4.3})$$

is an universal elastic modulus E^*, which can be expressed through the most important mechanical indices of the material and is a constant.

The ratio

$$\frac{\varepsilon_y}{\varepsilon_{lin}} = 1.6 \pm 0.1 \qquad (\text{A4.4})$$

indicates that the plastic deformation of polymeric glass begins when the current strain deviates from the linear response 1.6 times.

We will especially emphasize that the discovered universal regularities are not specific for polymeric glasses.

The deformation behaviour of

- low molecular weight organic, inorganic and metallic glasses;
- a number of metallic materials and single crystals;
- model systems, for example, a planar system of disks connected by non-valent interactions

satisfies the unified curve (Fig. A4.1) and the universal relations (A4.1)–(A4.4.) (see Section 4.1). In other words, the proposed unification of deformation behaviour is applicable not only to certain classes of materials, but to plastic bodies as a whole. Exceptions are materials that undergo phase transformations in the process of deformation – crystallizable polymers, complex alloys, mixtures and composites.

The physical meaning of this unification, which provides a universal description of the deformation process, irrespective of the temperature-velocity test conditions, is obvious. The constancy of the relations of mechanical parameters (expressions (A4.1)–(A4.4.)) implies an identical change in these characteristics with a change in temperature and strain rate. The stability of the profile of the unified deformation curve (Fig. A4.1) is explained by the fact that the relaxation effects that determine the development of deformation are hidden in the normalizing parameters σ_y and ε_y. In this connection, let us consider in more detail the temperature–rate dependences of these characteristics.

Figure A.4.3 shows the experimental temperature dependences of $\sigma_{f.e.}$ for a number of carbon-chain polymer glasses [60].

Replacement of the absolute deformation temperature T_{def} by the relative temperature $\Delta T = T_g - T_{def}$ (the difference between the glass transition temperature T_g and the deformation temperature T_{def}) leads to a unified temperature dependence of $\sigma_{f.e.}$ (Fig. A4.4). The value of ΔT varied in two ways: for an individual polymer with a fixed T_g, due to a change in T_{def}, and in the case of a fixed T_{def}, due to a change in T_g of the polymer during plasticization or copolymerization. Similar unified dependencies were obtained for the yiel strain ε_y, and also for the elastic modulus E_0 [15, 122, 123].

The results shown in Fig. A4.4 indicate that at deformation temperatures that are equidistant from the glass transition temperature, i.e. in appropriate states, polymeric glasses of different chemical structure are characterized by the same mechanical properties. This means that the change in the chemical structure determines only the change in the glass transition temperature of the polymer, leaving the complex of the mechanical properties of the material unchanged (in the corresponding states).

Thus, the polymeric glasses, as well as other plastic bodies, have a strictly defined similarity of the physico-mechanical behaviour, expressed in the existence of a unified deformation curve (Fig. A4.1) and universal numerical relationships (A4.1)–(A4.4.). Such similarity takes the relaxation character of the process of plastic

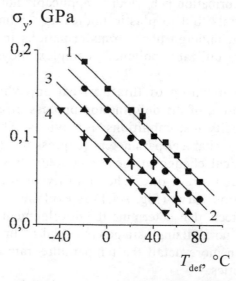

Fig. A4.3. Temperature dependences of the yeile stress σ_y for poly(methyl methacrylate) (1), poly(vinyl chloride) (2), polystyrene (3) and methyl methacrylate copolymer with lauryl methacrylate of composition 85/15 (4). Uniaxial compression. Strain rate: 1.7×10^{-4} s^{-1}.

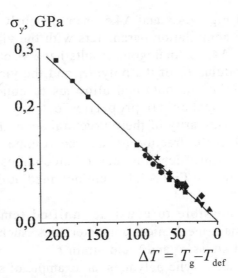

Fig. A4.4. The dependence of σ_y on the difference between the glass transition temperature T_g and the deformation temperature T_{def} for carbon-chain polymer glasses. Uniaxial compression. Strain rate: 1.7×10^{-4} s^{-1}.

deformation beyond the brackets and reduces its regularities to a system of correlations that are resistant to the chemical structure of the material and to the temperature and time regimes of exposure.

To answer the question cocnerning the nature of this similarity let us consider the results of structural studies obtained by positron annihilation. This method is widely used to study the evolution of free volume in polymers under various physical and physicochemical influences [65–69]. The fixed experimental parameters (long-lived components of the spectra – the lifetime τ_L and the intensity of positronium annihilation rate I_L), at least qualitatively, characterize the effective size of the fluctuation pores and their concentration, respectively. The product of the annihilation characteristics $\tau_L \times I_L$ has the meaning of the fluctuation free volume.

Figure A4.5 shows the dependence of the annihilation parameters of polymers of different chemical structure on the difference in the glass transition temperature and the test temperature $T_g - T_{test}$. These dependences reflect the universal nature of the effect of temperature on these characteristics. At test temperatures that are equidistant from the glass transition temperature, the annihilation spectra parameters characterizing the average hole size, hole concentration, and hence the free volume fraction in glassy polymers of different chemical nature are the same, indicating a similarity of their structural state, expressed in terms of the free volume.

Comparison of Figs. A4.4 and A4.5 unambiguously testifies to the correlation of the annihilation parameters with the yield stress σ_y of the material (Fig. A4.6). Analogous results have also been obtained for the elastic modulus E_0 of the polymer and the yield strain ε_y.

Thus, the correlations obtained allow us to conclude that the similarity of the mechanical properties of polymer glasses is determined by the similarity of their structural state, the quantitative criterion of which is the fraction of the free volume.

Unified temperature dependences of mechanical parameters, similar to that shown in Fig. A4.4, were obtained at different strain rates.

This made it possible to construct unified temperature–rate diagrams of the main mechanical characteristics, such as the elastic modulus E_0, yield stress σ_y and yield strain ε_y.

For vinyl and methacrylic polymers, an example of such a diagram is shown in Fig. A.4.7.

The experimentally obtained unified temperature–rate diagrams in the coordinates *(parameter)* = $f(\Delta T, \ln V)$ allow to unite universally the mechanical parameters of polymer glasses (E_0, σ_y, ε) with the relative deformation temperature ΔT and strain rate V:

$$(parameter) = A\Delta T \ln\left[\frac{V}{V_0}\right], \tag{A4.5}$$

where A is a coefficient describing the change in the mechanical characteristic with a change in ΔT and the strain rate for unity.

For the studied carbon-chain polymers [15,60], the temperature-rate dependence of σ_y is characterized by the value of A, equal to 0.0675 ± 0.003 MPa/deg. For other classes of polymers, for example, for polycarbonates, this parameter has a different value, but it is also constant. In other words, the coefficient A is a characteristic of the class of polymer glasses. In all cases, V_0 is a constant of the order of $10^{-12}-10^{-13}$ s^{-1}.

Summarizing, we conclude that the variety of relaxation effects responsible for the development of plastic deformation of glassy polymers can be taken into account by normalizing the current values of stress and strain to relaxing characteristics, namely, $\sigma_{f.e.}$ and $\varepsilon_{f.e.}$. The result of this procedure is a unified deformation curve (Fig. A4.1), the profile and quantitative parameters of which are resistant to changes in the temperature–time regimes of the action. Unification of the effect of temperature and velocity on the

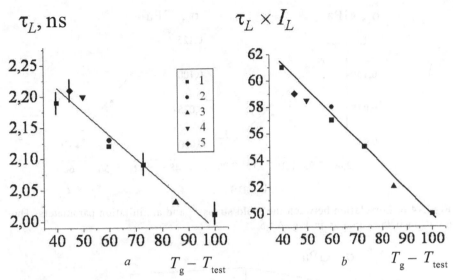

Fig. A4.5. Dependences of the lifetime τ_L (*a*) and the product of the lifetime on the annihilation intensity $\tau_L \times I_L$ (*b*) of positronium on the difference between the glass transition temperature T_g and the test temperature T_{test} for poly(methyl methacrylate) (1), polystyrene (2), methyl methacrylate copolymers with octyl methacrylate compositions 95/5 (3) and 80/20 (4) and a copolymer of methyl methacrylate with a lauryl methacrylate of composition 85/15 (5).

normalizing characteristics is achieved, in turn, by introducing a scale of relative temperatures ΔT (Fig. A4.7).

This approach has a pronounced predictive character and serves as a basis for rapid analysis of the physicomechanical behavior of new representatives of a certain class of polymer glasses without conducting mechanical tests.

The algorithm for such an analysis includes the following steps:

1. independent determination of the glass transition temperature of the new polymer, for example, using differential scanning calorimetry, which requires a small amount of sample;

2. an estimation of the values of σ_y and ε_y on the diagram (*parameter*) $= f(\Delta T, \ln V)$ (Fig. A4.7) for the temperature and strain rate of interest to the researcher;

3. substitution of these characteristics into a unified deformation curve (Fig. A4.1) and reproduction of the real deformation curve in the given temperature–rate conditions of deformation.

It is important to emphasize that the proposed methodology is completely based on experimental results and does not require the

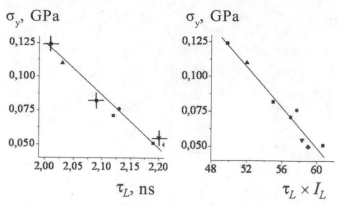

Fig. A4.6. Correlation between the yield stress σ_y and annihilation parameters. The notation is the same as in Fig. A4.5.

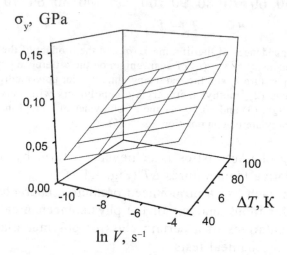

Fig. A4.7. Unified temperature–rate diagram of the yield stress σ_y for carbon-chain polymer glasses. Uniaxial compression

use of structural models, deterministic deformation mechanisms, representations of elementary deformation carriers, etc.

References

1. High-molecular compounds, ed. Zezin A.B., Moscow: Yurait, 2016, 340 p.
2. Govariker V.R., et al., Polymer science, New Delhi, Wiley Eastern Ltd., 1986.
3. Ward I.V., Sweeney, J., An Introduction to the Mechanical Properties of Solid Polymers, Chichester: John Willey & Sons Ltd., 2004, 382 p.
4. Bower D.I., An Introduction to Polymer Physics, Cambridge: University Press, 2002, 444 p.
5. Shaw M.T., MacKnight W.J., Introduction to Polymer Viscoelasticity, Hoboken, New Jersey: John Willey & Sons Ltd., 2005, 316 p.
6. Perepechko I.I., Introduction to the physics of polymers, Moscow: Khimiya, 1978, 311 p.
7. Arzhakov M.S., High-molecular compounds. Glossary of terms of concepts and definitions, Montreal: Accent Graphics Communications, 2016, 174 p.
8. Eyring, H.J., J. Chem. Phys., 1936, **4**, 283.
9. Krausz A.S., Eyring H.J., Deformation Kinetics, New York: Wiley-Interscience, 1976, 412 p.
10. Alexandrov A.P., Proceedings of the 1st and 2nd conferences on High-molecular compounds, Moscow: USSR Academy of Sciences, 1945, 45.
11. Reiner M., Physics Today, January, 1964, p. 62.
12. Arzhakov M.S., et al., Intern. J. Polymeric Mater., 2000, **47**, 149.
13. Ferry J.D., Viscoelastic properties of polymers, New York, John Wiley & Sons, 1940, 641 p.
14. Lukovkin G.M., et al., Dokl. Akad. Nauk, 2003, **391**, 4, 500.
15. Arzhakov M.S., et al., Introduction to the unified analysis of physical properties of substances and materials, Montreal: Accent Graphics Communications, 2017, 87 sec.
16. McCrum N., et al., Anelastic and Dielectric Effects in Polymeric Solids, London: Wiley, 1967, 617 p.
17. Transitions and relaxation phenomena in polymers, ed. Boyer R., Moscow: Mir, 1968, 384 p.
18. Bartenev G.M., Structure and relaxation properties of elastomers, Moscow: Khimiya, 1979, 288 p.
19. Bartenev G.M., Sanditov D.S., Relaxation processes in vitreous polymers, Novosibirsk: Nauka, 1986, 238 p.
20. Bershtein V.A., Egorov V.M.,, Differential scanning calorimetry in the physicochemistry of polymers, Leningrad: Khimiya, 1990, 248p.
21. Bartenev G.M., Barteneva A.G., Relaxation properties of polymers, Moscow: Khimiya, 1992, 382 p.
22. Heijboer J., Static and Dynamic Properties of the Polymeric Solid State, London: Reidel Publ. Co., 1982, pp. 197.

23. Johari G., Goldstein M., J. Chem. Phys., 1970, **53**, 6, 2372.
24. Hayler L., Goldstein M., J. Chem. Phys., 1977, **66**, 11, 4736.
25. Johari G., Polymer, 1986, **27**, 6, 866.
26. Bershtein V.A., et al., Dokl. Akad. Nauk USSR, 1983, **269**, 3, 627.
27. Bershtein V.A., Egorov VM, Vysokomol. Soed., A, 1985, **27**, 11, 2440.
28. Bershtein V.A., et al., Polymer Bull., 1983, **9**, 2, 98.
29. Glasstone S., et al., Theory of Rate Processes, McGraw-Hill, New York, 1941, 310 p..
30. Vinogradov G.V., Malkin A.Ya., Rheology of polymers, Moscow: Khimiya, 1977, 438 p.
31. Gotlib Yu.Ya., FTT, 1964, **6**, 10, 2938.
32. Gotlib Yu.Ya., J. Polym. Sci., Polym. Symp., 1968, **16**, 3365.
33. Gibbs J., DiMarzio E., J. Chem. Phys., 1958, **28**, 3, 373.
34. Adam J., Gibbs J., J. Chem. Phys., 1965, **43**, 1, 139.
35. Bueche E.J., Chem. Phys., 1956, **24**, 2, 418.
36. Boyer, R.J., Polym. Sci., Polym. Symp., 1975, **50**, 189.
37. Boyer R., 0Polymer, 1976, **17**, 11, 996.
38. Boyer, R.J., Macromol. Sci., Phys., 1980, **18**, 5, 461.
39. Boyer R.J., Macromol. Sci., Phys., 1980, **18**, 5, 563.
40. Gillham J., Boyer R.J., Macromol. Sci., Phys., 1977, **13**, 3, 497.
41. Enns G., Boyer R., Polym. Preprints, 1977, **18**, 462.
42. Lobanov A.M., Frenkel S.Ya., Vysokomolek. Soed., A, 1980, **22**, 5, 1045.
43. Hatakeyama T. J., Macromol. Sci., Phys., 1982, **21**, 2, 299.
44. Bershtein V.A., et al., Vysokomolek. Soed., 1978, **20**, 12, 2681.
45. Boyer R., J. Appl. Polym. Sci., 1986, **32**, 3, 4075.
46. Privalko V.P., Molecular structure and properties of polymers, Leningrad: Khimiya, 1986, 238 p.
47. Shakhparonov M.I., Introduction to the modern theory of solutions, Moscow: Vysshaya shkola, 1976, 296 p.
48. Tager A.A., Physicochemistry of Polymers, Moscow: Nauchnyi Mir, 2007, 576 p.
49. Taraoke I., Polymer Solutions. An Introduction to Physical Properties, New York: City University of New York, 2004, 360 p.
50. Stadnicki S., et al., J. Appl. Polym. Sci., 1976, **20**, 5, 1245.
51. Denny L., et al., J. Polym. Sci., Polym. Symp., 1984, 71, 39.
52. Tutov I.I., Kostrykina G.I., Chemistry and physics of polymers, Moscow: Khimiya, 1989, 432 p.
53. Barshteyn R.S., et al., Plasticizers for polymers, Moscow: Khimiya, 1982, 200 p.
54. Kozlov P.V., Papkov S.P., Physicochemical basis of plasticization of polymers, Moscow: Khimya, 1982, 224 p.
55. Liu Y., et al., Macromolecules, 1990, **23**, 968.
56. Vrentas J.S., et al., Macromolecules, 1988, **21**, 1470.
57. Robeson L.M., Faucher, J.A., J. Polym. Sci., Polym. Lett., 1969, **7**, 35.
58. Zhurkov S,N,, Dokl. Akad. Nauk SSSR, 1945, **47**, 7, 393.
59. Kargin V.A., Malinsky. Yu.M., Dokl. Akad. Nauka SSSR, 1950, **73**, 5, 967.
60. Arzhakov M.S., Thesis for a Doctoral Degree in Chemistry, 2004, Moscow: Moscow State University, 48 p.
61. Arzhakov M.S., et al., J. Polymeric Mater., 1998, **39**, 319.
62. Structural and Mechanical Behaviour of Polymer Glasses, Arzhakov M.S., et al., Commack, New York: Nova Science Publishers, 1997, 275 p.
63. Macquenn R.C., Granata R.D., J. Polym. Sci., 1993, **B31**, 971.

64. Sanchez V., et al., J. Appl. Polym. Sci., 1995, **56**, 779.
65. Kevdina I.B., et al., Vysokolmol. Soed., B, 1995, **37**, 4, 703.
66. Kevdina I.B., et al., Intern. J. Polymeric Mater., 1998, **39**, 313.
67. Shantarovich V.P., et al., Polymer Science, A, 1999, **41**, 7, 739.
68. Arzhakov M.S., et al., Intern. J. Polymeric Mater., 2000, **47**, 169.
69. Shantarovich V.P., et al., Materials Science Forum, 2001, **363-365**, 352.
70. Racich J.L., Koutsky J.A., J. Appl. Polym. Sci., 1976, **20**, 2111.
71. Bair H.E., et al., Macromolecules, 1972, **5**, 114.
72. Neki K., Geil P.H., J. Macromol. Sci., 1973, **B8**, 295.
73. Marikhin V.A., Myasnikova L.P., Supramolecular structure of polymers, Leningrad: Khimiya, 1977, 238 p.
74. Bershtein V.A., et al., Vysokomolek. Soed., A, 1985, **27**, 4, 771.
75. Egorov E.A., Vysokomolek. Soed., 1983, **25**, 4, 693.
76. Stamm M., et al., Faraday Disc. Chem. Soc., 1979, 68, 263.
77. Guttman C., DiMarzio E., Macromolecules, 1982, **15**, 2, 525.
78. Marikhin V.A., et al., Vysokomolek. Soed., A, 1986, **28**, 9, 1983.
79. Illers K., Europ. Polym. J., 1974, **10**, 911.
80. Selikhova V.I., et al., Vysokomolek. Soed., A, 1977, **19**, 4, 759.
81. Strobl G.R., The Physics of Polymers, Springer, 2007, 518 p.
82. Cowie J.M.G., Arrighi V., Polymers: Chemistry and Physics of Modern Materials, CRC Press, 2007, 520 p.
83. Young R.J., Lovell, P.A., Introduction to Polymers, CRC Press, 2011, 688 p.
84. Gul' V.E., Kuleznev V.N., Structure and mechanical properties of polymers, Moscow: Labyrinth, 1994, 367 p.
85. Kuleznev V.N., Shershnev V.A., Chemistry and physics of polymers, Moscow: Koloss, 2014, 368 p.
86. The Physics of Glassy Polymers, Haward R.N., Young R.L., Eds., 1997, Springer-Scince + Business Media, B.V., 508 p.
87. Robertson R.E., J. Chem. Phys., 1966, **44**, 3950.
88. Sanditov D.S., Bainova A.B., Fizika i khimiya stekla, 2004, **30**, 2, 153.
89. Sanditov D.S., et al., Fizika i khimiya stekla, 1996, **22**, 6, 683.
90. Sanditov D.S., et al., Fizika i khimiya stekla, 2000, **26**, 3, 323.
91. Sanditov B.D., Mantatov V.V., Vysokomolek. Soed. B, 1991, **32**, 2, 119.
92. Salamatina O.B.,et al., J. Therm. Anal., 1992, 38, 1271.
93. Oleinik E.F., et al., Polym. Adv. Tech., 1995, 6, 1.
94. Oleinik E.F., et al., Vysokomolek. Soed., A, 1993, **35**, 11, 1819.
95. Perez J., Polymer, 1988, 29, 483.
96. Perez J., et al., Rev. Phys. Appl., 1988, 23, 125.
97. Perez J., Rev. Phys. Appl., 1986, 21, 93.
98. Spaepen F., Acta Met., 1977, **25**, 3, 407.
99. Egami T., Vitek V., J. Non-Crystalline Solids, 1984, **62**, 4, 499.
100. Bowden P.B., Raha S., Phil. Mag., 1974, 29, 149.
101. Escaig B., Polym. Eng. Sci., 1984, July, **24**, 10, 737.
102. Escaig B., in: Plastic Deformation of Amorphous and Semicrystalline Materials, Escaig B., G'Sell C., Eds., Les Editions de Physique, Publ. Les Ulis, France, 1982, 187.
103. Cavrot,J. et al., Mater. Sci., 1978, 36, 95.
104. Haussy J., et al., J. Polym. Sci., Polym. Phys. Ed., 1980, 18, 311.
105. Hasan O.A., Boyce M.C., Polym. Sci. Eng., 1995, 35, 331.
106. Argon A.S., Phil. Mag., 1973, 28, 839.

107. Haward R.N., in: Molecular Behavior and the Development of Polymeric Materials, Ledwith A. and North A.M., Eds., 1974, London: Chapman and Hall, 444.
108. G'Sell C., Gopez A.J., J.Mater. Sci., 1985, 20, 3462.
109. Oleinik E.F., Adv. Polym. Sci., 1986, 80, 49.
110. Kozey V.V., Kumar S., J. Mater. Res., 1994, 9, 2717.
111. Wu J.B.C., Li J.C.M., J. Mater. Sci., 1976, 11, 434.
112. Li J.C.M., in: Plastic Deformation of Amorphous and Semicrystalline Materials, Escaig B., G'Sell C., Eds., Les Editions de Physique, Publ. Les Ulis, France, 1982, 359.
113. Ratner S.B., Brokhin Yu.I., Dokl. Akad. Nauk SSSR, 1969, 188, 4, 807.
114. Ratner S.B., Yartsev V.P., Physical mechanics of plastics, Moscow: Khimiya, 1992, 320 p.
115. Urzhumtsev Yu.S., Mechanics of Polymers, 1975, 1, 66.
116. Golman A.Ya., et al., Mechanics of Polymers, 1975, 3, 400.
117. Kovriga V.V., et al., Plast. massy, 1973, 4, 60.
118. Lohr J.J., Trans. Soc. Rheology, 1965, 9, 1, 65.
119. Kozlov G.V., Sanditov D.S., Anharmonic effects and physicomechanical properties of polymers, Novosibirsk: Nauka, 1994, 264 p.
120. Sanditov D.S., Fizika i khimiya stekla, 1991, 17, 4, 535.
121. Arzhakov M.S., et al., Deformatsiya i razrushenie materialov, 2005, 7, 2.
122. Lukovkin G.M., et al., Deformatsiya i razrushenie materialov, 2006, 6, 18.
123. Arzhakov M.S., et al., Materialovedenie, 2010, 7, 53.
124. Arzhakov M.S., et al., Dokl. Akad. Nauk, 1999, 369, 5, 629.
125. Arzhakov M.S., et al., Dokl. Akad. Nauk, 2000, 371, 4, 484.
126. Lukovkin G.M., et al., Dokl. Akad. Nauk, 2000, 373, 1, 56.
127. Glezer A.M., Molotilov B.V., Structure and mechanical properties of amorphous alloys, Moscow: Metallurgiya, 1992, 208 p.
128. Rapidly Quenched Metals, in: Proc. 5th Intern. Conf. on Rapidly Quenched Metals, Germany, eds. Warlimont H., Steeb S., 1984, 1112 p.
129. Agafonov Yu.V., et al., Physics of Classical Disordered Systems, Ulan-Ude: Buryat State University, 2000, 233 p.
130. Sanditov D.S., Fizika i khimiya stekla, 1989, 15, 4, 513.
131. Sanditov D.S., et al., Fizika i khimiya stekla, 1999, 25, 4, 416.
132. Berlin A.A., et al., J. Phys.: Condens. Matter, 1999, 11, 4583.
133. Kireev V.V., High-molecular compounds, Moscow: Yurait, 2013. 602 p.
134. Semchikov, Yu.D., High-Molecular Compounds, Moscow: Akademiya, 2005, 612 p.
135. Arzhakov M.S., Polymer Yearbook, 1998, 15, 179.
136. Godowsky Yu.K., Thermophysics of Polymers, Moscow: Khimiya, 1982, 280 p.
137. Volynsky A.L., Bakeev N.F., The role of surface phenomena in the structural and mechanical behavior of solid polymers, Moscow: Fizmatlit, 2014, 536 p.
138. Arzhakov S.A., et al., Vysokomolek. Soed., A, 1973, 15, 5, 119.
139. Bershtein V.A., et al., Vysokomolek. Soed., 1978, 20, 10, 2278.
140. Bershtein V.A., et al., FTT, 1981, 23, 6, 1611.
141. Bershtein V.A., et al., FTT, 1976, 18, 10, 3017.
142. Anishchuk T.A., et al., Vysokomolek. Soed., A, 1981, 23, 5, 963.
143. Arzhakov S.A., Thesis for obtaining the scientific degree of Doctor of Chemical Sciences, 1975, Moscow: NIFKhI.
144. Arzhakov M.S., et al., New Polymeric Materials, 1996, 5, 1, 43.
145. Bowden P.B., in: The Physics of Glassy Polymers, Haward R.N., Ed., 1973, John

Wiley, New York, Ch. 5.
146. Struik L., Polymer, 1987, **28**, 1, 57.
147. Bershtein V.A., Egorov V.M., FTT, 1984, **26**, 7, 1987.
148. Wendorff J., Proc. 4th Intern. Conf. Phys. Non-Cryst. Solids, Aedermannsdorf, 1977, 94.
149. Fischer E., Proc. 4th Intern. Conf. Phys. Non-Cryst. Solids, Aedermannsdorf, 1977, 43.
150. Peschanskaya N.N.,et al. Vysokomolek. Soed., A, 1985, **27**, 7, 1513.
151. Bessonov M.I., et al., Polyimides – the class of thermostable polymers, Leningrad: Nauka, 1983, 328 p.
152. Friedel J., Dislocations, Oxford, New York, Pergamon Press, 1964, 512 p.
153. Vladimirov V.I., Romanov A.E., Disclinations in Crystals, Leningrad: Nauka, 1986, 222 p.
154. Argon A.S., Phys. Chem. of Solids, 1962, **43**, 10, 945.
155. Argon A.S., Kuo H.Y., Mater. Sci. Eng., 1979, **39**, 1, 101.
156. Argon A.S., Shi L.T., Phil. Mag., A, 1982, **46**, 2, 275.
157. Bakai A.S., Polycluster amorphous bodies, Moscow: Energoatomizdat, 1987, 193 p.
158. Bakai A.S., Mater. Sci. Forum, 1993, 123–125, 145.
159. Bakai A.S., et al., Problems of Atomic Science and Technology, 2003, 3, 151.
160. Kamalova D.I., Proc. European Polymer Congress, Moscow, 2005, o.6.4.3.
161. Kamalova D.I., Thesis for obtaining the scientific degree of Doctor of Physical and Mathematical Sciences, 2007, Kazan: KSU, 248 p.
162. Schwarz O., et al., Processing of plastics, Moscow: Professiya, 2005, 315 p.
163. Kryzhanovskiy V.K., et al., Manufacture of products from polymeric materials, Moscow: Profession, 2004, 464 p.
164. Vlasov S.V., et al., Basics of plastics processing technology, Moscow: Khimiya, 2004, 600 p.
165. Sheryshev M.A., Technology of Polymer Processing, Moscow: Yurait, 2017, 302 p.
166. Shtarkman B.P., Fundamentals of the development of thermoplastic polymer materials, Nizhny Novgorod: Nizhny Novgorod Humanitarian Center, 2004, 328 p.
167. Baronin G.S., et al., Processing of polymers in the solid phase, Tambov: Publishing house of Tambov State Technical University, 2005, 61 p.
168. Arzhakov M.S., Arzhakov S.A., Intern. J. Polymeric Mater., 1997, **36**, 197.
169. Arzhakov M.S., Arzhakov S.A., Intern. J. Polymeric Mater., 1998, **40**, 1.
170. Arzhakov M.S., Lukovkin G.M., Arzhakov S.A., Dokl. Akad. Nauk, 2002, **382**, 1, 62.
171. Arzhakov M.S., et al., Deformatsiya i raztushenie materialov. 2009, 12, 12.
172. Rostiashvili V.G., et al., Glass Polymerization, Leningrad: Khimiya, 1987, 201 p.
173. Irzhak V.I., Architecture of polymers, Moscow: Nauka. 2012. 368 p.
174. Rao K.J., Rao C.N.R., Mater. Res. Bull., 1982, **13**, 5, 137.
175. Klinger M.I., Usp. Fiz. Nauk, 1987, **146**, 105.
176. Perepechko I.I., Acoustic methods for the study of polymers, Moscow: Khimiya, 1973, 296 p.
177. Belousov V.N., et al., Dokl. Akad. Nauk, 1993, **328**, 6, 706.
178. Belousov V.N., et al., Dokl. Akad. Nauk, 1990, **313**, 3, 630.
179. Belousov V.N., et al., Dokl. Akad. Nauk, 1985, **280**, 5, 1140.
180. Kozlov G.V., Novikov V.U., Usp. Fiz. Nauk., 2001, **171**, 7, 717.
181. Kozlov G.V., Zaikov,G.E., Structure of the Polymer Amorphous State, Leiden: Brill Academic Publishers, 2004, 465 p.
182. Bashorov M.T., et al., Nanostructures and properties of amorphous glassy polymers, Moscow: Russian Academy of Chemical Technology named after D.I. Mendeleev,

201, 269p.

183. Aloev V.Z., Kozlov G.V., Physics of Orientation Phenomena in Polymer Materials, Nal'chik: Polygrafservis, 2002, 285 p.

184. Shogenov V.N., Kozlov G.V., Fractal clusters in the physico-chemistry of polymers, Nal'chik: Polygrafservis, 2002, 267 p.

185. Dolbin I.V., et al., Structural stabilization of polymers, Moscow: Academy of Natural Sciences, 2007, 328 p.

186. Kozlov G.V., Aloev V.Z., Theory of Percolation in the Physicochemistry of Polymers, Nal'chik: Polygrafservis, 2005, 147 p.

187. Mikitaev A.K., Kozlov G.V., Fractal Mechanics of Polymer Materials, Nal'chik: Kabardino-Balkaria University, 2008, 311p.

188. Dyachkov A.I., Thesis for obtaining the scientific degree of Doctor of Chemical Sciences, 1986, Moscow: Lomonosov Moscow State University. 256 s.

Index

C

crystallization
  isothermal crystallization 64

D

differential scanning calorimetry 30, 37, 48, 52, 54, 55, 56, 67, 68, 71, 72, 90, 92, 150

E

energy
  Gibbs energy 59, 60, 61
  latent energy 35, 89, 90, 91, 92, 93
equation
  Arrhenius equation 31, 32
  Eyring relationship 34
  Falcher–Vogel-Tamman equation 106
  Kolmogorov–Avrami equation 65
  Thomson–Gibbs equation 68
  Williams–Landel–Ferry equation 27, 28, 29, 106

F

fibril 83, 84

I

Interstructural plastification 50
Intrastructural plastification 46

K

kinetic units 10, 11, 12, 14, 15, 16, 17, 18, 20, 24, 28, 30, 35, 39, 41, 55, 57, 69, 72, 78, 85, 104, 111, 122, 128, 130, 131

L

law